ADVANCED LEVEL

Practical Work for

PHYSICS

Practical Work for

PHYSICS

Chris Mee
Mike Crundell

Hodder & Stoughton

A MEMBER OF THE HODDER HEADLINE GROUP

Acknowledgements

The publishers would like to thank the following individuals, institutions and companies for permission to reproduce photographs in this book. Every effort has been made to trace ownership of copyright. The publishers are happy to make arrangements with any copyright holder whom it has not been possible to contact.

Andrew Lambert 1, 18, 31, 32 (both), 33
Philip Harris Education 38 (both), 39 (both), 40, 41, 45, 46, 47 (all three)
PhotoDisc 2

The illustrations were drawn by Jeff Edwards and Multiplex Techniques Ltd.

Orders: please contact Bookpoint Ltd, 130 Milton Park, Abingdon, Oxon OX14 4SB. Telephone: (44) 01235 827720, Fax: (44) 01235 400454. Lines are open from 9.00 – 6.00, Monday to Saturday, with a 24 hour message answering service. You can also order through our website www.hodderheadline.co.uk

A catalogue record for this title is available from The British Library

ISBN 0 340 78245 5

First published 2001
Impression number   10 9 8 7 6 5 4 3
Year   2007 2006 2005 2004 2003

Cover photo from Digital Stock.
Typeset by Multiplex Techniques Ltd.
Printed in Great Britain for Hodder & Stoughton Educational, a division of Hodder Headline Ltd, 338 Euston Road, London NW1 3BH by J.W. Arrowsmith, Bristol.

CONTENTS

PREFACE

The new Specifications (syllabuses) for A-level Physics published by the five Awarding Bodies in England, Wales and Northern Ireland, for first teaching from September 2000, emphasise Experiment and Investigation. Between 15% and 20% of the total marks at each stage of the assessment, that is, in both the AS and A2 years, must be allocated to practical work. The assessment is primarily skill-based, and must apply whether the Specification includes teacher-assessed coursework, an externally-set practical test, or both. The skills to be assessed are defined in detail in the QCA Physics Criteria. They include:

- *Devising and planning experimental activities*
- *Demonstrating practical techniques*
- *Observing, measuring and recording*
- *Interpreting, explaining and evaluating results.*

Some of the important synoptic assessment in the A2 year of the course may be achieved through Experiment and Investigation, particularly through questions on planning and design.

This book is focused on the four practical skills defined by the QCA Physics Criteria; it does not set out to be a book of instructions for a number of basic experiments. Instead, it commences with a chapter on data collection and recording, emphasising points of good practice. The second chapter deals with the processing of the collected data, and its presentation in tables and graphs. The third chapter summarises the experimental methods likely to be encountered in school and college laboratories for the measurement of the SI base quantities length, mass, time, temperature and current. The fourth chapter encourages the adoption of a critical attitude to numerical and graphical results, and introduces students to the evaluation of experimental procedures. The final chapter is directed towards teachers as much as students, and gives many examples of the way in which experimental activities can be aligned with assessment.

Recent International Syllabuses in Physics have the same broad assessment objectives as the new United Kingdom Specifications. The book is appropriate to any course of study in which practical work is assessed, or where good experimental technique is to be encouraged.

The book can be used as a stand-alone text. However, it is presented in a similar style to our theory textbook *AS/A2 Physics*, with worked examples and simple questions on each topic.

Chris Mee
Mike Crundell
March 2001

INTRODUCTION

Why spend valuable time in the Physics laboratory? Why does the AS/A2 course you are following demand that you acquire experimental and investigative skills? Why is a substantial proportion of the marks for the Specification (or syllabus) you are taking reserved for the assessment of these skills, either through coursework or through a formal practical test? Interactive computer simulations of many experiments are readily available. What will you gain by doing the experiment yourself?

A-level students experimenting with a fine-beam tube

The aims of a course in Practical Physics are as follows:

1 to demonstrate phenomena and laws (that is, to support the theories of Physics),

2 to test hypotheses (to find out if theories are correct),

3 to measure physical quantities (although it must be allowed that, for most of the quantities you will be asked to measure, you could more easily look them up in a book),

4 to develop manipulative skills (learn how to use measuring devices),

5 to provide training in the use of apparatus (to make sure that you don't damage expensive equipment),

6 to communicate the results of the experiment or investigation to others (because your fellow-students can benefit from your experience),

7 to plan, implement, analyse the evidence from, and evaluate experiments (that is, to complete the requirements of your Specification or syllabus).

To see a demonstration of a physics experiment, or to do one for yourself, is much more worthwhile than reading a description of it. For example, when you do an experiment to investigate the extension of a spring you will find out for yourself that Hooke's law is only obeyed over a limited range of forces and extensions. Although you might have seen a graph in your textbook which illustrates this very point, it is much more dramatic to plot the experimental points yourself and note the way in which a straight-line graph changes into a curve. You will see that, although you can measure the extension of the spring with an ordinary ruler, to improve on reliability you need to take precautions to avoid parallax error between the spring and the scale. If the spring breaks and the weight on the spring falls on your toes, you may also come to appreciate that certain safety precautions are sensible! These points, which may be passed over in a few minutes when reading a textbook, are learnt more thoroughly, and made much more interesting, by spending an hour or so in the laboratory.

The objective of providing training in the use of apparatus is very important. In your practical course you may need to use only simple instruments, such as rulers, vernier calipers, balances, ammeters and cathode-ray oscilloscopes. If you continue to study science beyond A-level you will use a much wider range of far more complex instruments. How will learning how to use a micrometer screw gauge help you to operate an X-ray machine or an electron microscope? In your course you will learn an attitude of mind, and acquire confidence, in the use of *any* scientific instrument. You will come to understand the importance of following detailed instructions, and you will appreciate the capabilities and limitations of the instrument. This attitude and confidence will be of the utmost value in your career.

The aim of teaching you to communicate your results is equally vital. The popular image of a scientist (and particularly of a physicist) is of a person working alone in a laboratory, surrounded by complex equipment, and eventually emerging with a vital formula such as $E = mc^2$. Nothing could be further from the truth. Science thrives on the interaction between scientists;

Animated discussion at a physics conference

it is one of the most social of occupations. Nowadays, experimental physicists hardly ever work in isolation. Many projects are so ambitious that teams of scientists, often based in laboratories in different countries, work together on complementary aspects of the problem. The most prestigious university and industrial laboratories provide areas where workers can meet over coffee to discuss their results and pool ideas in the most informal way. Of course, results are also communicated through seminars, lectures and conferences, and through the publication of reports and papers in learned journals. It is equally important that you are able to relate your science knowledge to friends who may not be studying Physics. You will start to learn this approach in your AS/A2 course. You should find it an enjoyable and stimulating part of your studies, and something you could never achieve through even the most interactive computer simulation. An important part of your assessment is how you present your results, conclusions and evaluations in written coursework and in sessions with your teachers.

In class, you are taught the theory of Physics. Phenomena are explained in terms of simple models involving idealised situations. (In how many problems have you been told to 'neglect air resistance' or 'assume ideal gas behaviour'?) From these simple models we develop a hypothesis, or learn a theory, and are told about deductions which can be made from the theory. However, the hypothesis, theory and deductions are of no value until they are tested. The test comes in the laboratory, when you do an experiment. This is why it is important to study experimental physics.

This book is directed towards four basic skills of experiment and investigation in Physics. They are:

1 Planning
2 Implementing
3 Analysing evidence and drawing conclusions
4 Evaluating evidence and procedures

These are the skills which must be assessed in any Specification leading to the award of a GCE AS or A-level certificate in Physics. Many other syllabuses test the same skills (but may not use exactly the same headings). The book does not treat the skills in the order presented above. We start with a chapter on data collection and recording, in which basic ideas of accuracy and precision are stated. Common errors in reading instruments are summarised. A consistent way of recording results in tabular form is suggested. The second chapter takes you through the correct way to process and present your experimental results. In the third chapter, the ideas involved when planning an experiment are put forward. You will learn how to select instruments to measure the quantities you will come across in experimental work. The fourth chapter goes on to discuss the evaluation of experimental results and procedures. Throughout, the importance of communication skills is emphasised. The final chapter (which is directed towards teachers as much as to students) describes the assessment of the four skills which are the focus of the course, and gives examples of tasks, experiments and investigations which may be used in the laboratory.

By following a course in practical Physics, you will begin to appreciate that it is not easy to design an experiment which tests a theory conclusively. You will learn that difficulties occur at almost every stage of an investigation. More importantly, you may find that you have acquired the experience and knowledge which will help you to overcome these difficulties. The most thrilling aspect of all is that you may think a problem is dry and theoretical in the classroom, but find that it comes to life when you test it by experiment in the laboratory. You will find a new enjoyment in discussing your results and sharing your experiences with your teachers and fellow-students.

CHAPTER ONE

Data collection and recording

Physics is sometimes called the science of measurement. Certainly, without observation and measurement, Physics would not exist. In order that theories may be developed and then tested, it is necessary to make measurements that are both precise and accurate. In this chapter we shall look at precision and accuracy. Also, we shall discuss how to record these measurements and the data obtained from them.

1.1 Accuracy and precision

Accuracy is the degree to which a measurement approaches the 'true value'. Accuracy depends on the equipment used, the skill of the experimenter and the techniques involved. Reducing error or uncertainty in a measurement improves its accuracy. Precision is that part of accuracy which is within the control of the experimenter. The experimenter may choose different measuring instruments and may use them with different levels of skill, thus affecting the precision of measurement.

If we want to measure the diameter of a steel sphere or a marble, we could use either a metre rule, a vernier caliper or a micrometer screw gauge. The choice of measuring instrument would depend on the precision to which we want the measurement to be made. For example, the metre rule could be used to measure to the nearest millimetre, the vernier caliper to the nearest tenth of a millimetre and the micrometer screw gauge to the nearest one-hundredth of a millimetre. We could show the readings as follows:

metre rule: 1.2 ± 0.1 cm
vernier caliper: 1.21 ± 0.01 cm
micrometer screw gauge: 1.212 ± 0.001 cm

The degree of precision to which the measurement has been made increases as we move from the metre rule to the vernier caliper and finally to the micrometer screw gauge. Note that the number of significant figures quoted for the measurement increases as the precision increases. In fact, the number of significant figures in a measurement gives an indication of the precision of the measurement. Precision is partly to do with the accuracy of an observation or measurement.

1.2 Uncertainty

In Section 1.1 each of the measurements was shown with its precision. For example, using the metre rule, the measurement of the diameter of the marble is 1.2 cm with a precision of 0.1 cm. In reality precision is not the only factor affecting the accuracy of the measurement. The total range within which the measurement is likely to lie is known as its **uncertainty**. For example, a measurement of 46.0 ± 0.5 cm implies that the most likely value is 46.0 cm, but it could be as low as 45.5 cm or as high as 46.5 cm. The uncertainty in the measurement is ± 0.5 cm or ± (0.5/46) × 100% ≈ ± 1%.

Uncertainty gives an indication as to the range of values within which the measurement is likely to lie.

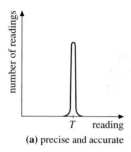

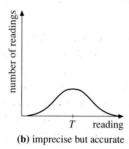

(a) precise and accurate (b) imprecise but accurate

Figure 1.1

When a measurement is repeated many times with a precise instrument, the readings are all close together, as shown in Figure 1.1a. Using a measuring instrument with less precision means that there would be a greater spread of readings, as shown in Figure 1.1b, resulting in greater uncertainty. A reading may be very precise but it need not be accurate. **Accuracy** is concerned with how close a reading is to its true value. For example, a micrometer screw gauge may be precise to ± 0.001 cm but, if there is a large zero error, then the readings from the scale for the diameter of a sphere or marble would not be accurate. The distinction between precision and accuracy is illustrated in Figure 1.2. On each of the graphs, the value T is the true value of the quantity.

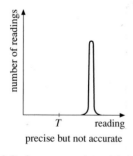

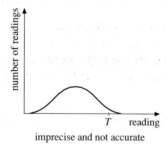

precise but not accurate   imprecise and not accurate

Figure 1.2 *Accuracy and precision*

Note: It is important that, when writing down measurements, the number of significant figures of the measurement indicates its uncertainty. Some examples of uncertainty are given on the next page.

Instrument	Uncertainty	Typical reading
Stopwatch with 0.2 s divisions	± 0.1 s	16.2 s
Thermometer with 1 deg C intervals	± 0.5 deg C	22.5 °C
Ammeter with 0.1 A divisions	± 0.05 A	2.15 A

Note that a particular temperature is shown as a number with the unit °C. A temperature interval is shown as a number with the unit deg C. However, many people nowadays use the unit °C for both a particular temperature and a temperature interval.

It should be remembered that the uncertainty in a reading is not wholly confined to the reading of its scale or to the skill of the experimenter. Any measuring instrument has a built-in uncertainty. For example, a metal metre rule expands as its temperature rises. At only one temperature will readings of the scale be precise. At all other temperatures, there will be an uncertainty due to the expansion of the scale. Knowing by how much the rule expands would enable this uncertainty to be removed and hence to improve precision.

Manufacturers of digital meters quote the uncertainty for each meter. For example, a digital voltmeter may be quoted as ± 1% ± 2 digits. The ± 1% applies to the total reading shown on the scale and the ± 2 digits is the uncertainty in the final figure of the display. This means that the uncertainty in a reading of 4.00 V would be $(\pm 4.00 \times 1/100) \pm 0.02 = 0.06$ V. This uncertainty would be added to any further uncertainty due to a fluctuating reading.

Note that some people refer to the uncertainty in a measurement as being its error. This is not strictly true. Error would imply that a mistake has been made. There is no mistake in taking the measurement, but there is always some doubt or some uncertainty as to its value.

Now it's your turn

1. A large number of precise readings for the diameter D of a wire are made using a micrometer screw gauge. The gauge has a zero error $+E$, which means that all readings are too large. Sketch a distribution curve of the number of readings plotted against the measured value of the diameter.

2. The manufacturer of a digital ammeter quotes its uncertainty as ± 1.5 % ± 2 digits.

    (a) Determine the uncertainty in a constant reading of 2.64 A.
    (b) The meter is used to measure the current from a d.c. power supply. The current is found to fluctuate randomly between 1.58 A and 2.04 A. Determine the most likely value of the current, with its uncertainty.

    Ans (a) ± 0.06 A (b) (2.01 ± 0.05) A

1.3 Choice of instruments

The precision of an instrument required for a particular measurement is related to the measurement being made. Obviously, if the diameter of a hair is being measured, a high precision micrometer screw gauge is required rather than a metre rule. Similarly, a galvanometer should be used to measure currents of the order of a few milliamperes, rather than an ammeter. Choice is often fairly obvious where single measurements are being made, but care has to be taken where two readings are subtracted. Consider the following example.

The distance of a lens from a fixed point is measured using a metre rule. The distance is 95.2 cm (see Figure 1.3). The lens is now moved closer to the fixed point and the new distance is 93.7 cm. How far has the lens moved? The answer is obviously 95.2 – 93.7 = 1.5 cm but how accurate is the measurement?

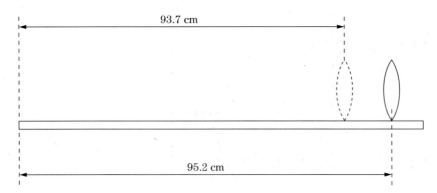

93.7 cm

95.2 cm

Figure 1.3

We have seen that the uncertainty in each measurement using a metre rule is, optimistically, ± 1 mm ($\pm \frac{1}{2}$ mm at the zero end of the rule plus $\pm \frac{1}{2}$ mm when finding the position of the centre of the lens). This means that each separate measurement of length has an uncertainty of about $(1/940 \times 100)\%$, about 0.1%. That appears to be good! However, the uncertainty in the distance moved is ± 2 mm (both distances have an uncertainty, and these uncertainties add up) so the uncertainty in calculating the change in position is $\pm (2/15 \times 100)\% \approx \pm 13\%$. This uncertainty is, quite clearly, unacceptable. Another means to measure the distance moved must be devised in order to reduce the uncertainty.

During your AS/A2 course, you will meet with many different measuring instruments. You must learn to recognise which instrument is most appropriate for particular measurements. A stopwatch may be suitable for measuring the period of oscillation of a pendulum, but you would have difficulty using it to find the time taken for a stone to fall vertically from rest through a distance of 1 m. Choice of appropriate instruments is likely to be examined when you are planning experiments.

Now it's your turn

1. Suggest appropriate instruments for the measurement of:

    (a) the discharge current of a capacitor (of the order of 10^{-6} A)
    (b) the time for a feather to fall in air through a distance of about 40 cm
    (c) the time for a ball to fall vertically through a distance of about 40 cm
    (d) the length of a pendulum having a period of about 1 s
    (e) the temperature of some water as it cools to room temperature
    (f) the temperature of a roaring Bunsen flame
    (g) the weight of 20 small glass beads
    (h) the weight of a house brick.

2. The diameter of a ball is measured using a metre rule and a set square, as illustrated in Figure 1.4. The readings on the rule are 16.8 cm and 20.4 cm. Each reading has an uncertainty of ± 1 mm.

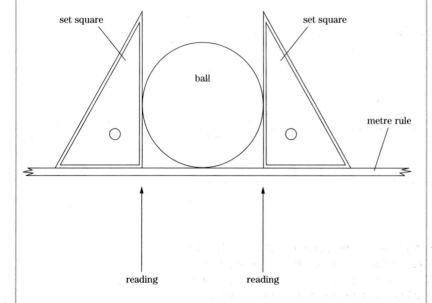

Figure 1.4 *Measuring the diameter of a ball*

Calculate, for the diameter of the ball:

(a) its actual uncertainty,
(b) its percentage uncertainty.

Suggest an alternative but more precise method by which the diameter could be measured.

Ans (a) ± 0.2 mm, (b) ± 6 %

1.4 Systematic and random uncertainty (error)

Not only is the choice of instrument important so that any measurement is made with acceptable precision, but the techniques of measurement must also optimise accuracy. That is, your experimental technique must reduce as far as possible any uncertainties in readings. These uncertainties may be classed as either systematic or random.

Systematic uncertainty (error)

A systematic uncertainty will result in all readings being either above or below the accepted value. This uncertainty cannot be eliminated by repeating readings and then averaging. Instead, systematic uncertainty can be reduced only by improving experimental techniques. Examples of systematic uncertainty are:

- zero error on an instrument (the scale reading is not zero before measurements are taken – see Figure 1.5)

- a parallax error (the scale is not viewed normally when taking a reading – see Figure 1.6)

- wrongly calibrated scale

- reaction time of experimenter.

Zero errors

Where appropriate, instruments should be checked for any zero error. If there is no zero error, then a note should be made to this effect in the written account of your experiment. Where there is a zero error, the meter should be adjusted to zero or, if this not possible, the zero error should be noted in your account and all recorded readings should then be adjusted.

Note: Remember to record all readings as they are taken. Do not allow for zero error 'in your head' and then write down the adjusted value.

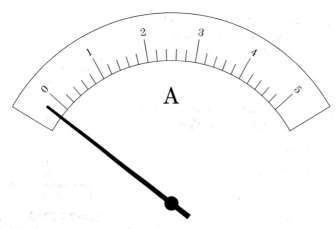

Figure 1.5 *A zero error on a scale*

For example,

zero error on micrometer = – 0.003 cm

reading on micrometer for diameter of wire = 0.146 cm

corrected reading for diameter of wire = 0.149 cm

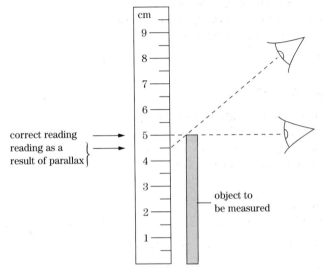

Figure 1.6 *Parallax*

Parallax errors

To reduce parallax errors, always

1   have the scale as close as possible to the pointer

2   view the scale normally (i.e. so your line of sight is perpendicular to the scale, see Figure 1.7).

The scales on some meters are fitted with a small plane mirror to help to view the scale normally. When the image of the pointer in the mirror is hidden behind the pointer, then the scale is being viewed correctly and the reading can be taken (see Figure 1.8). Similarly, a small plane mirror may be used to assist with reading a metre rule, as illustrated in Figure 1.9.

An alternative to the use of a plane mirror is a set square placed along a metre rule, as illustrated in Figure 1.10. The set square is run along the edge of the rule until it reaches the point to be measured (see also Figure 1.4).

Wrongly calibrated scales

Generally, in school physics laboratories we assume measuring devices are correctly calibrated and we are not expected to check the calibration in an experiment. However, it may be of interest to check the calibration of:

● ammeters; connect several in series in a circuit

● voltmeters; connect several in parallel in a circuit

● rulers (especially plastic rulers); lay them side by side

● thermometers; put them in well-stirred water.

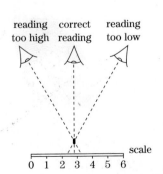

Figure 1.7 *Parallax error*

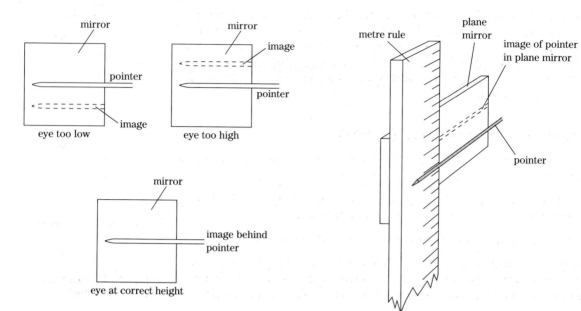

Figure 1.8 Use of a plane mirror

Figure 1.9 Using a plane mirror with a metre rule

These checks will not enable you to say which of the various instruments are calibrated correctly, but they will show up variations between them.

Reaction times

Where timings have to be made manually, it has to be accepted that there will be a delay between the experimenter observing an event and starting a stopwatch. This delay is known as the reaction time. Where possible, electronic switching of timers is advisable. The effects of reaction time may be reduced by:

- starting and stopping the watch to the same stimulus, i.e. both visual, or both hearing a sound

- making the time interval between starting and stopping the watch as large as possible.

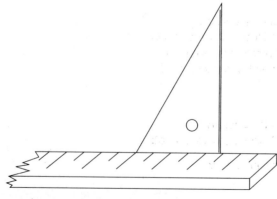

Figure 1.10 Using a set square with a metre rule

For example, when determining the period of a pendulum, the bob should be viewed as it passes through the equilibrium position. You should become accustomed to the rhythm of the oscillations by counting down to zero (3, 2, 1, 0, ...) as the bob passes the equilibrium position and then start the watch at the zero count. Count twenty oscillations, then stop the watch. Repeat the procedure three times.

Random uncertainty (error)

Random uncertainty results in readings being scattered around the accepted value. Random uncertainty may be reduced by

- repeating a reading and averaging (see 'Reaction times' where the time for twenty oscillations of the pendulum is repeated)

- plotting a graph and drawing a best fit line to the points.

Note: When readings are repeated, you must always record all the readings and then show the average. Do not merely write down the average. Some examples of good practice to reduce random uncertainty are given below.

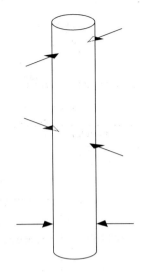

Figure 1.11 Measuring the diameter of a wire

Timing oscillations – see 'Reaction times'
1 Check for zero error on the stopwatch.

2 Count sufficient oscillations so that the time is greater than about ten seconds (twenty seconds is even better as this further reduces percentage uncertainty).

3 Stop and then restart the oscillations – there may be a systematic error due to the wrong mode of oscillation (e.g. oscillations not in one plane for a simple pendulum).

4 Repeat the timing of the oscillations. Ideally, three timings should be made.

Measuring the diameter of a wire
1 Check the micrometer for any zero error.

2 Measure the diameter at several positions along the length of the wire (to allow for tapering).

3 Measurements along the length of the wire should be taken spirally (to check for circular cross-section), see Figure 1.11.

Use of plane mirror or set square when reading a scale
These techniques are discussed on pages 11 and 12. Remember that, when writing your account of the experiment, the use of these items should be mentioned. To save time, a sketch diagram is very helpful (see Figures 1.9 and 1.10).

Second set of readings
It is not always appropriate to repeat a reading and then average. Examples of cases where this is not suitable include the measurement of:

1 current and potential difference values to determine the V/I characteristic of an electrical component

2 load and extension values to determine the force/extension or force/displacement characteristic of a spring or cantilever.

In situations like these, particular values of one of the variables should be chosen, and then the other variable should be measured as the former is increased and then decreased. That is:

1 For particular values of potential difference, measure the current as the potential difference is first increased and then decreased. Average the values of current.

2 For particular values of load, measure the extension as the load is first increased and then decreased. Average the values of extension.

Drawing a graph

Graphs provide a particularly useful means of averaging readings. Ideally, the quantities plotted on the x- and y-axes should be chosen so as to yield a straight line (see Chapter 2). There will be an uncertainty in each plotted point, but drawing the best fit line effectively takes an average. Moreover, studying a graph with a best fit line enables the following:

- any wayward point can be identified and checked by taking further measurements.

- a weighted average can be taken. That is, those points close to the anticipated line may be given more emphasis than those further away.

Note: Where a wayward point is identified and a second set of measurements is made, the original measurements and the plotted point should not be erased. Instead, both the erroneous point and its replacement should be clearly identified.

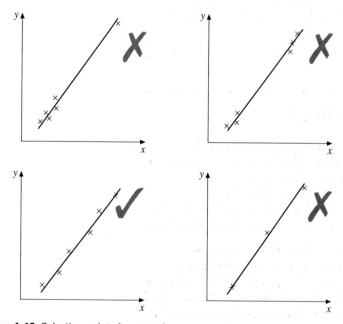

Figure 1.12 Selecting points for a graph

It should be remembered that a graph is only as good as the data used to plot it. In order that a graph should be reliable it is important that, when you are planning your experiment, you should:

- have at least six sets of readings,

- ensure that the points on the graph are spread evenly along the line (see Figure 1.12).

Too few points mean that too much emphasis is placed on individual readings which may not be reliable. Similarly, if points are not evenly distributed along the line, then any isolated point is given undue emphasis. Experimental design for graphical work and the analysis of graphs are both discussed in Chapter 2.

Now it's your turn

Suggest what measurements would have to be made, and the procedure to be adopted in the following experiments:

(a) Determine the variation with voltage of the current in a 12 V filament lamp.

(b) Take measurements to investigate how the size of the image produced by a convex lens depends on image distance.

(c) Obtain values of the time taken for a ball to roll different distances down an incline.

(d) Determine the variation with time of the height of the column of water draining out of a burette.

(e) Investigate the dependence of period of oscillation of a bifilar suspension on the length L of the suspension. (A bifilar pendulum is shown in Figure 5.9).

1.6 Recording of readings

It has been mentioned already that it is important to record all your readings. Indeed, the instructions in some examinations are '... *to plan your work and to record all readings as they are taken. A rough copy of readings is not to be made*'. In all assignments, marks are lost unless all readings are shown. There is no point in recording an average value unless all the readings from which the average has been calculated are also shown.

It follows that, in some experiments, there will be a large number of readings and these must be presented in a form which is easily read and understood. Generally, the quickest and easiest way to present readings is in the form of a table. A table also allows for the processed data (the results of calculations based on readings) to be shown with the readings themselves. As an example, we can consider a cantilever (see Figure 1.13).

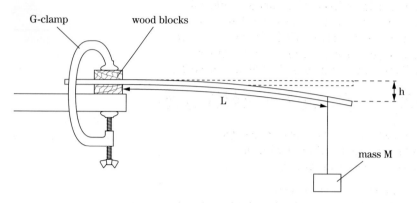

Figure 1.13

Measurements are to be made of the depression of the end of the cantilever, and also the period of oscillation of the cantilever, for different loads. A graph of the depression of the cantilever is to be plotted against (period)^2. Figure 1.14 is a table of the measurements and the processed data.

Load /N	reading at end of rule/cm		average reading /cm	depression h/cm	time for 20 oscillations/s			period T/s	T^2/s^2
	Load increasing	Load decreasing			1	2	3		
O	3.2	3.2	3.2	–	–	–	–	–	–
2.5	11.1	11.3	11.2	8.0	13.4	13.6	13.3	0.672	0.451
3.5	14.7	14.7	14.7	11.5	16.4	16.0	15.8	0.803	0.645
4.5	18.8	19.0	18.9	15.7	17.8	17.6	18.0	0.890	0.792
5.5	20.5	20.8	20.7	17.5	20.2	20.4	19.6	1.00	1.00
6.5	24.2	23.8	24.0	20.8	21.4	21.8	21.6	1.08	1.17
7.5	28.4	28.4	28.4	25.2	24.4	23.8	23.8	1.20	1.44

Figure 1.14 *Table of results for a cantilever*

When constructing a table for readings and measurements (data), and calculated results (processed data), it is important to remember:

1 There should be a natural progression when moving from left to right. That is, raw data appears on the left hand side, and processed data on the right.

2 All columns must be headed showing the quantity and the unit in which it is measured. Columns should be headed in much the same way as graph axes are labelled (see Chapter 2).

3 The number of decimal places for data must be consistent, and must indicate the uncertainty in the reading (see Section 1.2).

4   The number of significant figures in processed data should reflect the
uncertainty in the raw data. If raw data are given to two significant
figures, then processed data should be to two or three significant figures.
The correct use of significant figures is discussed in more detail in
Chapter 2. For further detail on the construction of tables, see page 20.

Now it's your turn

A student has been asked to investigate how the resistance of a
thermistor varies with temperature. Figure 1.15 is a table of
measurements and processed data that he produces.

Identify as many examples of bad practice as you can.

potential difference /V	current/mA		average current I	resistance kΩ	temperature /C
	temperature increasing	temperature decreasing			
3.0	96	87	91.5	3.28	4.0
3.02	120	114	117	2.58	10
3.0	134	138	136	2.21	15
3.1	175	180	177	1.75	21
1.58	114	129	122	1.3	30
0.76	170	161	165	0.461	50

Figure 1.15

Ans: There are eight different examples of bad practice, each of which
may lead to a deduction of marks in an assessed piece of work.

Chapter 1 Summary

★   Accuracy is concerned with how close a reading is to its true
value.

★   Precision is the part of accuracy which can be controlled by
the experimenter.

★   Uncertainty gives an indication of the range of values within
which a measurement is likely to lie.

★   A systematic uncertainty (error) is often due to instrumental
causes, and results in all readings being above or below the
true value. It cannot be eliminated by averaging.

★   A random uncertainty (error) is due to scatter of readings
around the true value. It may be reduced by repeating a
reading and averaging, or by plotting a graph and taking a
best-fit line.

CHAPTER TWO

Data processing

In Chapter One we were concerned with the collection and recording of data. Once this data has been collected, it must be processed before any conclusions can be drawn from it. For example, there may be three readings of the time taken for ten oscillations of a pendulum. These three readings must be processed in order to find the period of oscillation. In this Chapter, we will look at the correct processing of data and the way in which it should be presented.

Timing the oscillations of a simple pendulum. What improvements could you make?

2.1 Significant figures in processed data

The number of places of decimals in recorded (or raw) data provides information as to the accuracy with which the measurements were made. For example, if the time for ten oscillations of a pendulum are measured to be 23.2 s, 23.8 s and 23.6 s, then it is reasonable to assume that each reading could be taken to the nearest 0.2 s (or, possibly, the nearest 0.1 s).

The value for the period of oscillation can then be found.

$$\text{period} = (23.2 + 23.8 + 23.6) \div 30$$

$$= 2.35333\ldots \text{ seconds.}$$

Having been careful to ensure that the number of decimal places in the raw data is sensible and has real meaning, then some thought should be given to the processed data.

In the example above, it would appear as if we can determine the period to the nearest 0.00001 s! This is, quite clearly, nonsense. Since the times for ten oscillations were each measured to the nearest tenth of a second, then it would seem reasonable that the time for one oscillation can be stated to the nearest hundredth of a second, i.e. 2.35 s. This is because one tenth of a second divided by ten is one hundreth of a second. Note that the raw data was given to three significant figures and the processed data is also to three significant figures. The general rule for giving an appropriate number of significant figures in processed data is:

> Processed data should be given to the same number of significant figures as raw data.

Example

Complete the following, giving the answers to an appropriate number of significant figures.

$L = 22$ cm,	$L^2 =$	Ans: **480 cm²**
$\theta = 15.4°$	$\tan \theta =$	Ans: **0.275**
$T = 7.46$ s	$\lg T =$	Ans: **0.873**
$T = 74.6$ s	$\lg T =$	Ans: **1.873**
$T = 746$ s	$\lg T =$	Ans: **2.873**

The abbreviation **lg** stands for '**logarithm to the base ten**'. Your calculator probably uses the abbreviation **log**. Be careful not to confuse **log** with **ln**. The abbreviation **ln** stands for '**logarithm to the base e**'.

Note that it appears at first sight that lg 74.6 and lg 746 have been quoted to too many significant figures. This is not the case. When taking logarithms, the integer in front of the decimal point in the logarithm gives the position of the decimal point in the actual number. That is,

$$7.46 = 7.46 \times 10^0, \quad \lg 7.46 = 0.873$$
$$74.6 = 7.46 \times 10^1, \quad \lg 74.6 = 1.873$$
$$746 = 7.46 \times 10^2, \quad \lg 746 = 2.873$$

When logarithms are taken, only the numbers to the right of the decimal point should be considered when deciding on the number of significant figures to be used.

It may well be necessary to give more significant figures than in the raw data when the result varies rapidly. For example,

$$\tan 85° = 11.4 \qquad \text{or} \qquad 67^3 = 300800$$
$$\tan 86° = 14.3 \qquad\qquad\qquad 68^3 = 314400$$
$$\tan 87° = 19.1 \qquad\qquad\qquad 69^3 = 328500$$

In each of these cases, the raw data was given to two significant figures. However, two significant figures for the processed data may not be appropriate since the changes in the function are so rapid. In examples such as these, you must use your discretion as to how many significant figures should be quoted. The fact remains that you must be consistent and not vary the number given for any one set of processed data.

Now it's your turn

Complete the following, giving the answers to an appropriate number of significant figures.

$$X = 7\text{cm}, \qquad X^2 = \rule{3cm}{0.4pt}$$
$$Y = 67.3°, \qquad \sin Y = \rule{3cm}{0.4pt}$$
$$L = 98, \qquad \lg L = \rule{3cm}{0.4pt}$$
$$A = 98, \qquad A^3 = \rule{3cm}{0.4pt}$$

Ans: 50cm^2, 0.923, 1.99, 941000

2.2 Presentation of processed data

Tables

In Chapter 1, the advantages of presenting data in the form of a table were discussed, along with the various conventions which should be observed.

Processed data should be included within the same table as the raw data.

Columns should be headed in the same way as for raw data. When constructing the table, there should be a natural progression from left to right. This means raw data should be in columns on the left of the table and processed data towards the right.

length d/cm	time for 20 oscillations /s	period T /s	lg (d/cm)	lg (T/s)
67.4	16.2 16.6 16.0	0.813	1.829	–0.0897
75.8	etc …			

In this example, note that each column is headed with a physical quantity, perhaps written out in full ('time for 20 oscillations'), or perhaps designated by a word and a symbol ('length d'). If the length had been defined as d in the text of the laboratory report, it would be satisfactory to head the column as 'd'. In each column, the unit of the quantity is included after an oblique slash ('d/cm'). This notation is continued in the columns giving the processed data. In particular, note that the logarithm columns are headed 'lg (d/cm)'. This is because the logarithm of the length d depends on the unit in which d is measured. If d is 67.4 cm, lg (d/cm) is lg 67.4 = 1.829. But if the same distance is measured in mm as 674 mm, lg (d/mm) is lg 674 = 2.829. It would be incorrect to head this column simply as 'lg d'.

Now it's your turn

Identify failures of good practice and any other errors in the table below.

distance travelled d/cm	time taken /s			lg (d/cm)	average time t/s	lg (t/s)
20.5	3.2	3	3.4	1.312	3.2	0.51
30	3.8	4.0	3.8	1.477	3.87	0.59
42.5	etc.....					

Ans: Better to transpose the third and fourth columns
first column: all data should be to 3 significant figures
second column: all data should be to 2 significant figures
fourth column: all processed data should be to 2 significant figures

Graphs

In many experiments, several sets of data for a pair of variables are collected. These sets of data are analysed in order to find a general relation between the two variables. Graphs provide a very good means by which a relation can be determined. Furthermore, by looking at the scatter of points on the graph of Figure 2.1, the reliability of the stated result for the relation may be estimated. More scatter indicates less reliability.

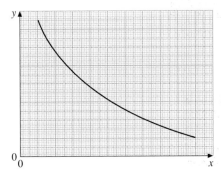

Figure 2.1 *As x increases, y decreases*

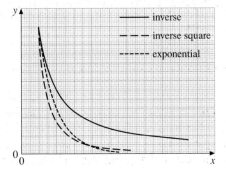

Figure 2.2 *Inverse, inverse square and exponential changes*

Plotting points on a graph that results in a curve may enable a general trend to be determined. In Figure 2.1, it is possible to say that as x increases, y decreases. We cannot say with any certainty whether the relationship is inverse, inverse square or exponential (see Figure 2.2).

The graph in which we can have most confidence is a straight line graph. Consequently, for most purposes we attempt to plot a graph that will result in a straight line.

Looking at the history of physics, there have been numerous occasions where important detail has been missed as a result of assuming a result and then attempting to prove it. Similarly, in your work in the laboratory, you may be asked to find the relation between the length l of a pendulum and the period T of its oscillations. You suspect that T depends on l and so you plot a graph of either T against $\sqrt{l}$ or T^2 against l. Both of these graphs should produce a straight line. What if a curve is produced? You can then say what it is not, but you can't say what it is!

It is always considered to be good practice *not* to assume the result but, rather, to design an experiment to determine the relation.

We will return to this in Chapter 3 when we look at the planning of experiments, but at present we do need to understand how to analyse a straight line graph.

Straight line graphs

If a variable y is plotted on the y-axis and a variable x is plotted on the x-axis as illustrated in Figure 2.3, then, if the graph is a straight line, the line may be represented by the equation

$$y = mx + c$$

m is the gradient of the graph and c is the intercept on the y-axis. Referring to Figure 2.3, the gradient m is given by

$$m = (y_2 - y_1) / (x_2 - x_1)$$

and the value of c is the value of y when $x = 0$.

Thus, any relation can be found completely if it is in the form of a straight line graph. The task is to ensure that any relation appears as a straight line.

Think about an experiment where you have been asked to find the relation between two variables, P and Q. There may be other variables, but these must be kept constant in this particular experiment. In this example, it may be assumed that the relation between P and Q is of the form

$$P = aQ^n$$

where a and n are constants. Taking logarithms of both sides of the equation,

$$\lg P = \lg a + n \lg Q$$

$$\text{or, } \lg P = n \lg Q + \lg a$$

Comparing this equation with the equation for a straight line, i.e.

$$y = mx + c$$

then it can be seen that plotting a graph of $\lg P$ on the y-axis against $\lg Q$ on the x-axis, a straight line will be produced. The gradient m will equal n and the intercept c will equal $\lg a$. The relation between P and Q can thus be found.

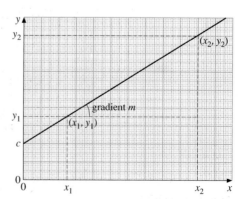

Figure 2.3 A straight line graph

Example

Quantities R and S are related by the expression

$$R = aS^n$$

Use the graph of Figure 2.4 to determine the relation between R and S.

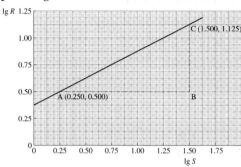

Figure 2.4 A lg-lg graph

Gradient $= CB / BA$

$\qquad = (1.125 - 0.500) / (1.500 - 0.250)$

$\qquad = 0.500$

Intercept $= 0.375$

Since $R = aS^n$, then $\lg R = n \lg S + \lg a$

n is the gradient and $\lg a$ is the intercept.

$n \qquad = 0.50$
$\lg a \quad = 0.375,$
$a \qquad = 10^{0.375} = 2.4$

The relation is $R = 2.4\ S^{0.50}$ or $R = 2.4\sqrt{S}.$

Now it's your turn

Quantities Y and X are related by the expression

$$Y = aX^n$$

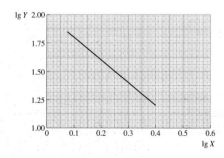

Figure 2.5

Use the graph of Fig. 2.5 to determine the relation between Y and X.

Ans: $Y = 100/X^2$

During your studies of physics you will come across exponential changes. For example, both radioactive decay and the discharge of a capacitor through a resistor follow an exponential variation. Such changes require a slightly different approach.

Consider the quantity Y which varies exponentially with x according to the relation

$$Y = Y_0 \, e^{-bx}$$

This could be, for example, the variation with time of the activity of a radioactive source.

Taking natural logarithms of this equation,

$$\ln Y = \ln Y_0 - bx$$

and, re-arranging, $\qquad \ln Y = -bx + \ln Y_0$

If we now compare this with the equation for a straight line,

i.e. $\qquad\qquad\qquad y = mx + c$

then it can be seen that if we plot $\ln Y$ on the y-axis and x on the x-axis, the gradient of the line will be equal to $-b$ and the intercept on the y-axis will be $\ln Y_0$. Hence the form of the exponential can be deduced.

Other graphs

We may not always wish to plot a straight line graph and, possibly, the relation we are investigating may not be of any of the forms discussed previously. For example, we may be investigating the variation with external resistance of the output power of a battery. This relation is shown in Figure 2.6. It can be seen that the output power increases to a maximum and then decreases. Obviously, Figure 2.6 is the most appropriate graph if we wish to determine the maximum output.

In Chapter 1 we discussed the choice and number of readings and in general, readings should be equally spaced. However, there is an exception here. In Figure 2.6 the important region is near the maximum, so extra points should be plotted near the maximum to enable a more precise value to be obtained.

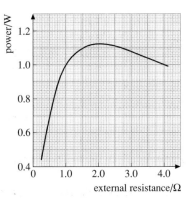

Figure 2.6 *Output power of a battery*

2.3 Plotting graphs

We have seen that graphs are a very useful means of presenting data. In particular, the advantages are:

1 The general trend of various sets of data may be seen.

2 Drawing a best-fit line to the plotted points enables average behaviour to be determined. Furthermore, the average is weighted. When drawing the best-fit line, points which do not follow the general trend of the line are not given as much attention as those close to the line. These points may even be ignored. However, it is good experimental practice to repeat such readings so as to check whether they are faulty.

3 Drawing a best-fit line enables the scatter of the points to be seen. This scatter enables the reliability of the experiment to be determined (see Chapter 4).

In order that graphs may be used to their full advantage, they must be drawn in the correct way. The remainder of this section describes how this should be done.

Choice of scales

The scale on an axis determines the distance along that line which represents one unit of measurement. For example, if we are plotting electrical power on the y-axis, then 1 cm could represent 1 W or 2 cm could represent 1 W, and so on, depending on the range of powers to be covered.

The correct choice of scale is important. Larger scales mean that points can be plotted with greater precision, but we do not want a scale which is awkward to read, for example, 3 cm representing 1 W.

Having examined the sheet of graph paper on which the graph is drawn, you should consider:

1 The scale on both axes. The area in which the points will be plotted should be at least one half of the sheet of graph paper. However, points should not appear as an 'extension' to the graph (see Figure 2.7).

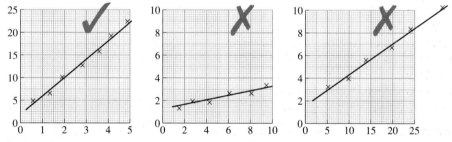

Figure 2.7 *Choosing a suitable scale (n.b. the axes have not been labelled)*

Remember that, where an intercept is to be found, the origin of the graph should be included. However, if this means that all the plotted points are in a very small area of the graph, then the origin should not be included. The intercept is then found by calculation (see pages 27 and 28).

2 The scale must be easy to read. Acceptable scales are 1:1, 1:2, 1:5, 1:10 etc. Unacceptable scales are 1:3, 2:3, 3:5 etc. (see Figure 2.8).

Drawing axes and scales

Each axis should be drawn on the graph grid as a thin straight line.

Values should be marked on the axes at regular intervals (either every large square or every other large square) and in the conventional directions. This means that on the x-axis, values increase to the right and on the y-axis, values increase 'upwards'. A common error is to leave a 'hole' in the scale. Some scales are shown in Figure 2.8.

Having marked the scale, the graph is of no value unless the axes are labelled. Axes should be labelled with the quantity and its unit in a similar way to the column headings for data (see Chapter 1). As an example, if an axis is used to plot speed in m s⁻¹, then it should be labelled as speed/m s⁻¹ or speed (m s⁻¹).

Note that, if logarithms are being plotted, then a logarithm is just a number. Consequently, the unit should be shown in brackets with the quantity. For example, if the logarithm to the base 10 of the period T of a pendulum, measured in seconds, is to be plotted, then the axis should be labelled as lg (T/s).

Plotting the points

All points should be plotted as accurately as possible, and certainly to within one half of a small square. When plotting the point, it should be shown as

either a diagonal cross   ✕

or a horizontal cross   +

or a dot within a circle   ⊙

It is important that, if you show graph points as dots, then you circle the dot. Otherwise, when the graph line is drawn, the line may go through the point and the point would be lost from view. Remember that you are likely to lose marks in examinations if you do not plot all points clearly.

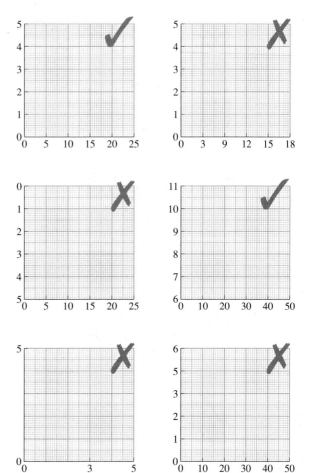

Figure 2.8 *Scales and axes (n.b. the axes have not been labelled)*

The best-fit line

The best-fit line should be a thin clear line without being 'lumpy' or 'hairy' (see Figure 2.9).

The plotted points should be scattered equally about the line. Do be careful not to force the line to pass through the origin. This failure to pass through the origin may well be due to a systematic error in measurement (see Chapter 1). One important feature of graphical analysis is that it enables some types of systematic error to be identified. Furthermore, by measuring the gradient of the graph, allowance is made for such errors.

A further warning! Some relationships may start off as straight lines but then tail off, for instance the V/I characteristic of a filament lamp. You must be prepared to draw such graphs as a straight line which then becomes a curve (see Figure 2.10).

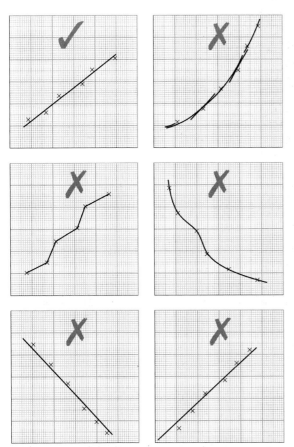

Figure 2.9 *Drawing the best-fit line (n.b. axes have not been labelled)*

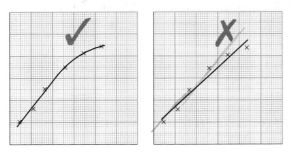

Figure 2.10 *Graph which is not totally linear (n.b. axes have not been labelled)*

 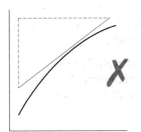 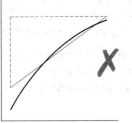

Measuring the gradient

The method by which the gradient of a graph can be calculated was shown on page 23. It is important, when drawing the triangle from which the gradient will be determined, that the hypotenuse is at least one half of the length of the line on the graph. If this is not done, then the coordinates used to calculate the gradient will be too close and consequently this will adversely affect the accuracy of the calculated gradient.

If the gradient at a point on a curve is required, then a tangent to the curve should be drawn at that point (see Figure 2.11). When drawing a tangent, it is helpful to use either a transparent ruler or a plane mirror. Viewing the line and its image in the plane mirror enables a line to be drawn which is symmetrically placed between the line and its image. The length of the tangent for calculation of the gradient must be about half the length of the curve.

Remember that the coordinates used when calculating the gradient must be for points lying on the straight line or tangent. They should not be values from the table of results, unless the values actually lie on the line.

Determining an intercept

Where the origin of the graph can be included then it is easy to read off the intercept. However, where inclusion of the origin would mean that all points are plotted in a small area of the graph, then it is not sensible to include the origin. The intercept must then be found by calculation, as in the next example.

Having found the gradient of the graph line, the coordinates of one point on the line can be read. If this gradient and the coordinates are substituted in to the equation for the relationship, the intercept can then be found.

Figure 2.11 *Drawing a tangent at a point on a curve*

Example

Consider the graph of Figure 2.12. The axes have not been labelled and should be thought of as numbers without units. The intercept on the y-axis is not 2.5.

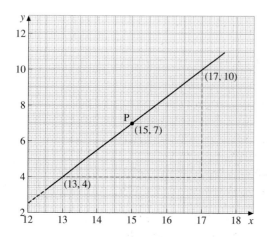

Figure 2.12 *Determination of an intercept*

Gradient $= m = \Delta y / \Delta x$

$\qquad = (10.0 - 4.0) / (17.0 - 13.0)$

$\qquad = 1.50$

For the point P, $x = 15.0$, $y = 7.0$ and the gradient of the line $= 1.50$.

Substituting into the equation of a straight line,

$$y = mx + c$$

then, $\qquad 7.0 = (1.50 \times 15.0) + c$

$\qquad c = -15.5$

The intercept on the y-axis is -15.5 and the equation of the line is

$$y = 1.50x - 15.5$$

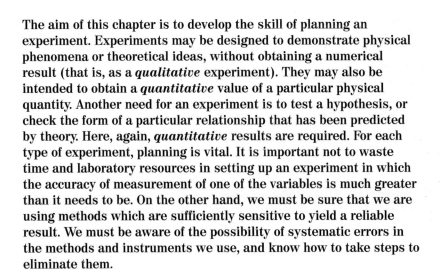

CHAPTER THREE

Planning

The aim of this chapter is to develop the skill of planning an experiment. Experiments may be designed to demonstrate physical phenomena or theoretical ideas, without obtaining a numerical result (that is, as a *qualitative* experiment). They may also be intended to obtain a *quantitative* value of a particular physical quantity. Another need for an experiment is to test a hypothesis, or check the form of a particular relationship that has been predicted by theory. Here, again, *quantitative* results are required. For each type of experiment, planning is vital. It is important not to waste time and laboratory resources in setting up an experiment in which the accuracy of measurement of one of the variables is much greater than it needs to be. On the other hand, we must be sure that we are using methods which are sufficiently sensitive to yield a reliable result. We must be aware of the possibility of systematic errors in the methods and instruments we use, and know how to take steps to eliminate them.

3.1 Methods and techniques

All experiments designed to obtain a quantitative result for a physical quantity involve measurements. These measurements must be of some combination of the base quantities length, mass, time, temperature and current. (To complete the list, we should include quantity of substance and luminous intensity, but these are not encountered in experimental work in Physics for the AS or A2 assessment.) In the following sections we shall look at the methods available for measuring the base quantities in a school or college laboratory. By understanding the principles of the available methods, we shall be able to make an informed decision about the choice of a particular technique, with respect to making the experiment as precise and reproducible as possible, and to avoiding sources of systematic error. For all of the quantities, the effective choice will be limited by what instruments are available in your laboratory. However, in one type of examination question, on planning and design, you may be asked to devise an experiment and draw on your theoretical rather than practical knowledge of various types of apparatus.

At AS/A2 level, students generally assume that the calibration of the instruments they use is correct. However, it is worth thinking about how to compare the calibration of one instrument against another, even if this is a check you will very seldom make. In a planning/design question, you might be asked to suggest a method of calibration. Generally, as has been mentioned on page 11, it is easy to *compare* the calibration of different instruments, but not so easy to determine which of two instruments giving different readings is correct.

In the sections that follow, we shall spend most time on methods for measuring length, simply because length-measuring instruments of several different types will be available in your laboratory.

Length

The metre rule

The simplest length-measuring instrument to be found in your laboratory is a metre (or half-metre) rule. It has the great advantages of being cheap, convenient and simple to use. A relatively unskilled student should have no difficulty in taking a reading with an uncertainty of 0.5 mm. However, you should be aware of three possible sources of error.

The first may arise if the end of the rule is worn, giving rise to a zero error (see Figure 3.1). For this reason, it is bad practice to place the zero end of the rule against one end of the object to be measured and to take the reading at the other end. You should place the object against the rule so that a reading is made at each end of the object. The length of the object is then obtained by subtraction of the two readings. A zero error like this is a *systematic* error, because it is involved every time a reading is taken from the zero end of the rule. In general, the zero reading of any instrument may be subject to an error. We shall meet this type of error again in the micrometer screw gauge, and in the ammeter.

The calibration of the metre rule may give rise to another systematic error because the markings are incorrect. Try comparing the 30 cm graduated length of one plastic rule with the same nominal length on another. You are quite likely to find a discrepancy of one or two millimetres. One of the reasons why wooden or plastic metre rules are cheap is that the manufacturer does not claim any great accuracy for the scale markings. The calibration may be checked by laying the rule alongside a more accurate rule, such as a steel 30 cm or metre rule, and noting any discrepancy. If you compare an engineer's steel rule with a plastic rule, you will see at once that the engraved marks on the steel rule are much finer than the impressed marks on the plastic. Of course, the extra care which has been taken in engraving the steel rule has to be paid for. A one-metre steel rule is many times more expensive than a plastic metre rule.

Another source of error with the metre rule is parallax error. If the object to be measured is not on the same level as the graduated surface of the rule, the angle at which the scale is viewed will affect the result as illustrated in Figure 3.2. This is a *random* error, because the angle of view may be different for different readings. It may be reduced by arranging the rule so that there is no gap between the scale and the object. Parallax error is also important in reading any instrument in which a needle moves over a scale. A rather sophisticated way of eliminating parallax error is to place a mirror alongside the scale on the meter rule. When the needle and scale are viewed

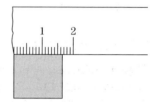

Figure 3.1 *Zero error with a metre rule*

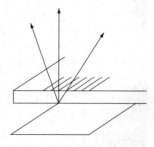

Figure 3.2 *Parallax error with a metre rule*

directly, the needle and its image in the mirror coincide. This ensures that the scale reading is always taken at the same viewing angle.

The smallest division on the metre rule is 1 mm. If you take precautions to avoid parallax error, you should be able to estimate a reading to about 0.5 mm. If you are measuring the length of an object by taking a reading at each end, the uncertainties add to give a total uncertainty of 1 mm. The range of the metre rule is from 1 mm to 1000 mm. To measure a length of more than 1 m with a metre rule will introduce a further uncertainty, of perhaps 1 mm or 2 mm, because of the difficulty of making a reference mark at the 1 m end of the rule and moving the rule so that the zero exactly corresponds with this reference. It is usually better to use a steel tape to measure lengths of more than 1 m.

The micrometer screw gauge

The type of micrometer screw gauge available in a school or college laboratory may be used to measure the dimensions of objects up to a maximum of about 50 mm. Measurements can easily be made with an uncertainty of 10 μm or less. The principle of the instrument is the magnification of linear motion using the circular motion of a screw. The instrument consists of a U-shaped piece of steel with a fixed, plane, end-piece A. Opposite this is a screw with a corresponding end-piece B. The position of the screw can be adjusted using the ratchet C which is connected to the thimble D. There are graduations along the barrel of the instrument (the bearing in which the screw turns), and round the circumference of the thimble. The purpose of the ratchet is to ensure that the same torque (that is, the amount of twist) is applied to the thimble for each reading. If this torque is exceeded, the ratchet slips. The object to be measured is placed in the jaws of the gauge between the end-pieces A and B, and B is screwed down on to the object, using the ratchet C, until the ratchet slips.

The screw advances exactly 1 mm for two revolutions. That is, the pitch of the screw is 0.5 mm or 500 μm. If you look at the graduations on the barrel

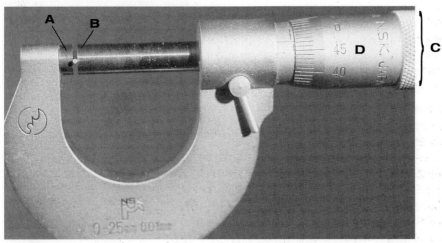

Micrometer screw gauge

of the screw bearing you will see that there are divisions every 0.5 mm. The reading on the barrel corresponds to the position of the edge of the thimble (Figure 3.3). When taking a reading it is important to check which half of the millimetre the edge of the barrel is in. The graduations round the circumference of the thimble run from 0 to 50. Each division corresponds to one-hundredth of a mm, or 10 µm. The reading on the thimble is added to the reading on the barrel. Thus, Figure 3.3 shows a reading of 9.5 mm on the barrel plus 0.36 mm on the thimble, 9.86 mm in total. You can easily read to the nearest division on the thimble, that is to the nearest 0.01 mm (10 µm).

The micrometer screw gauge is very likely to have a systematic zero error. Every time you use the gauge, you should check the zero error by moving face B so that it makes contact with face A. The screw must be tightened with the ratchet C, so that a reproducible zero is obtained. Then take the reading on the barrel and the thimble. This gives the zero error, which must be allowed for in all subsequent readings. Figure 3.4a shows a screw gauge with a zero error of + 0.12 mm. If this were the error which applied when the reading of 9.86 mm was obtained in Figure 3.3, the true length of the object would be 9.86 mm − 0.12 mm = 9.74 mm. Figure 3.4b shows a zero error of − 0.08 mm. In this case, the true length of the object in Figure 3.3 would be 9.86 mm + 0.08 mm = 9.94 mm.

In the case of a wooden or plastic metre rule, it is good practice to check the calibration against an engineer's steel rule, if one is available in your laboratory. It would be unusual to do the same with a micrometer screw gauge, but if there is doubt about the calibration of a particular gauge, it can be checked by measuring the dimensions of a series of gauge blocks. A gauge block is a rectangular steel block with faces which are accurately plane and parallel. The length of the gauge block is known to an uncertainty of less than 1 µm. However, not many school laboratories possess gauge blocks.

The vernier caliper

A vernier caliper is a versatile instrument for measuring the dimensions of an object, the diameter of a hole, or the depth of a hole. Its range is up to about 100 mm, and it can be read to 0.1 mm or 0.05 mm depending on the type of vernier with which it is fitted. It consists of a steel mm scale A with two reference posts at the zero mark. The sliding part B moves along the scale. The slider has the vernier scale engraved on it. The zero of the vernier corresponds with reference posts on the sliding part. One set of reference posts, those with the straight parts on the inside, is used like the jaws of a screw gauge: the object to be measured is placed between the jaws, and the sliding part B is moved along until the object is gripped tightly. A reading to the nearest mm is taken on the fixed scale, at the zero end of the vernier scale. The reading to a tenth of a mm is obtained by finding where a graduation of the vernier scale coincides with a graduation of the fixed scale. Figure 3.5 shows the scale of a vernier caliper giving a reading of 25.4 mm.

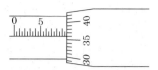
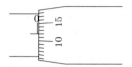
Figure 3.3 Screw gauge scale reading with a reading of 9.86 mm

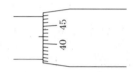
Figure 3.4a Zero error of +0.12 mm

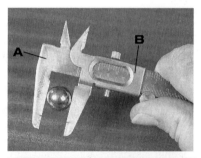
Figure 3.4b Zero error of −0.08 mm

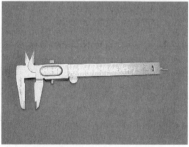
Vernier caliper

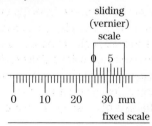
Figure 3.5 Vernier scale with a reading of 25.4 mm

Travelling microscope

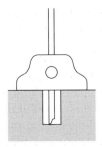

Figure 3.6 *Measurement of the depth of a blind hole*

The second set of jaws has the straight parts on the outside. These can be used to measure the diameter of a hole. The jaws are placed inside the hole and are moved apart until they are in contact with the edges of the hole. The scale and vernier can then be read.

A pin at the end of the sliding part of the caliper can be used to measure the depth of a hole, for example a hole which has been drilled in, but not right through, a wooden board. The end of the fixed scale is placed on the board, across the hole, and the pin moved into the hole until it reaches the bottom (see Figure 3.6). The reading of the scale and vernier gives the depth of the hole.

As with the micrometer screw gauge, the vernier caliper should be checked for a systematic zero error before taking a reading. The calibration can be checked using gauge blocks, but this precaution would be unusual in a school or college laboratory.

The travelling microscope

A specialised instrument for length measurement is the travelling microscope. This consists of a low-power microscope, with a magnification of about 10, mounted on a slider which is moved along rails by means of a screw. There is a fixed mm scale alongside the rails, and the carriage has a vernier or micrometer scale. The microscope is fitted with cross-wires. It can be focused on one feature of an object (for example, one edge), and the scale and vernier or micrometer reading taken. The microscope is then moved along the rails by turning the screw until the the cross-wires coincide with the other edge of the object. The difference between the scale readings gives the width of the object. Because results are always obtained as the difference between two readings, a zero error does not arise. However, because the microscope is driven along the rails by the screw and the scale is mounted on the frame, an error can arise because there is a certain amount of free play (backlash) in any screw drive. Thus, the microscope can be moved slightly without the scale reading showing any change. This becomes important if the direction of turning the screw is reversed. For this reason, the final adjustment of the slider to bring the cross-wires into coincidence with the chosen feature of the object should always be made so that the microscope is moving in the same direction.

The range of objects which can be measured depends on the length of the rails. Models in use in schools and colleges have rails up to 250 mm long.

Readings are to the nearest 0.1 mm or 0.05 mm, depending on the type of vernier or micrometer which has been fitted. Because measurements are by the difference of two readings, the uncertainty in a measurement is twice the uncertainty in a single reading.

Applications of the travelling microscope include measurement of the diameter of a thin tube, and of the separation of interference fringes in a two-slit experiment. In such an experiment, the fringe pattern may be cast on a ground-glass screen, and viewed with the microscope. Alternatively, the fringes may be viewed directly. In this case, a pin is placed in front of the microscope to determine the plane of viewing.

A feature of the travelling microscope is that the instrument does not make contact with the object to be measured. For the measurement of remote objects, a cathetometer, or travelling telescope, may be used (although it is very unlikely that this will be available in your laboratory).

When measuring the diameter of a wire, rod, or sphere with a screw gauge, caliper or travelling microscope, it is good practice to take readings spirally along its length, in case the object does not have a cross-section which is perfectly circular. Readings of the diameter of a rod or wire should be taken at intervals along the object, to allow for the possibility of taper.

Example

1   Figure 3.7a shows the scale of a micrometer screw gauge when the zero error is being checked, and Figure 3.7b shows the scale when the gauge is tightened on an object. What is the length of the object?

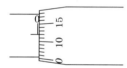   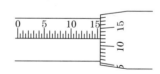

Figure 3.7a   *Figure 3.7b*

From Figure 3.7a, the zero error is + 0.12 mm. The reading in Figure 3.7b is 15.62 mm. The length of the object is thus (15.62 – 0.12) mm = **15.50 mm**.

2   Figure 3.8 shows the scale of a vernier caliper. What is the reading?

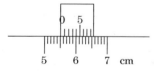

Figure 3.8

The zero of the vernier scale is between the 5.5 cm and 5.6 cm divisions of the fixed scale. There is coincidence between the seventh graduation of the vernier scale and one of the graduations of the fixed scale. The reading is thus **5.57 cm**, or **55.7 mm**.

Now it's your turn

1   Figure 3.9a and 3.9b show the scales of a micrometer screw gauge
    when the zero is being checked, and again when measuring the
    diameter of an object. What is the diameter?

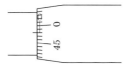

Figure 3.9a   **Figure 3.9b**

Ans: **12.52 mm**

2   Figs 3.10a and 3.10b show the vernier scales of a travelling
    microscope when it is focused on the opposite ends of the diameter
    of a glass tube. What is the diameter? What is the uncertainty in the
    diameter?

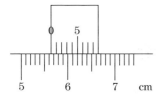   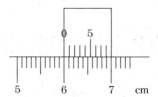

Figure 3.10a   **Figure 3.10b**

Ans: **3.5 mm; ± 0.2 mm**

Choice of method

A summary of the range and reading uncertainty of length-measurement
instruments is given in Figure 3.11.

Instrument	Range	Reading uncertainty	Notes
Metre rule	1 m	0.5 mm	check zero, calibration errors
Micrometer screw gauge	50 mm	0.01 mm	check zero error
Vernier caliper	100 mm	0.1 mm	versatile: inside and outside diameters, depth
Travelling microscope	250 mm	0.1 mm	no contact with object

Figure 3.11   *Typical range and uncertainty of readings taken with common
laboratory instruments*

In deciding which instrument to use in a particular experiment, you should consider first of all the nature of the length measurement you have to make. For example, if you need to find the diameter of a steel sphere, the micrometer screw gauge and caliper techniques are obvious candidates. You should then consider whether you need the greater accuracy of the micrometer. In a particular experiment, the uncertainty in the diameter of the sphere may be the dominant uncertainty (see Chapter 1), and in such a case the fact that the accuracy available with the screw gauge is ten times that for the vernier caliper will sway the argument. In an experiment which may last some time, you should also think about the resources of your laboratory. Is it sensible to tie up what may be one of only a small number of available screw gauges, when they may also be in demand by other students for other experiments? Sometimes, in design questions, you are asked to think about the cost of setting up an experiment. Here you should realise that a travelling microscope is by far the most expensive of the four instruments we have considered. We have also mentioned the difference in cost of a steel metre rule compared with wooden or plastic rules. It would be foolish economics to supply a steel rule for each of a number of students, when it would be perfectly adequate to provide each of them with a wooden rule and have one steel rule available in the laboratory for calibration purposes.

Application: measurement of pressure difference

A difference in pressure may be measured by comparing the heights of liquid in the two arms of a U-tube. Figure 3.12 shows a U-tube connected to a container of gas. The pressure above the liquid in tube A is atmospheric pressure p_{atm}. The pressure above the liquid in tube B, and hence the pressure of gas in the container, is p. The relation between the pressures is

$$p = p_{atm} + \Delta h \rho g$$

where Δh is the difference in vertical height between the levels of the liquid in the two arms of the tube, ρ is the density of the liquid, and g is the acceleration of free fall. To find the pressure of the gas, all we need to do is to measure the difference Δh between the liquid levels, assuming that p_{atm}, ρ (and g) are known.

In some laboratories, a U-tube mounted on a board to which a vertical millimetre scale is attached may be available. (This device is called a manometer.) If it is, it is a simple matter to measure Δh. If the manometer contains oil or water, the liquid in the tube will have a curved surface which is concave downwards (Figure 3.13a). This surface is called the *meniscus*. Use a set-square to find the reading on the vertical scale corresponding to the *bottom* of the meniscus in each side of the U-tube. The surface of the liquid in a manometer filled with mercury will be convex (Figure 3.13b), and in this case you should read to the *top* of the meniscus. Again, use a set-square to avoid parallax error. If a manometer is not available, you will have to arrange your own system of U-tube and scale, for example a half-metre rule. Make sure that the scale is clamped vertically (use a plumb line).

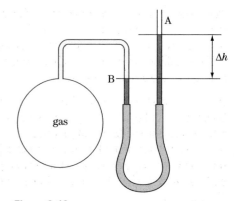

Figure 3.12

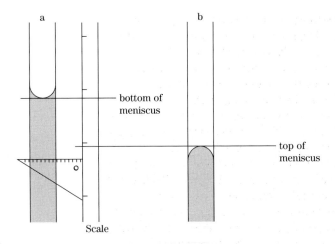

Figure 3.13

In some situations it is not possible to mount a vertical scale near the tube containing liquid. An example is an experiment to demonstrate the magnetic properties of solutions. A U-tube containing the solution is placed with one arm of the tube vertically between the poles of an electromagnet, and the other outside the region of the magnetic field. When the magnetic field is applied, the liquid in the tube between the pole-pieces may rise. To obtain quantitative results, the change in liquid level for a given applied magnetic flux density (field strength) is required. A good (although expensive) way of measuring the small change in the liquid level is to use a travelling telescope (cathetometer), or perhaps a travelling microscope (see page 33), if this can be brought sufficiently close to the tube. The cross-wires of the telescope or microscope should be aligned with the bottom of the liquid meniscus. If you use a travelling microscope or telescope for this type of measurement, remember that the image will be inverted. Thus, in this experiment, the *rise* of the liquid in the tube will be seen as a *fall* when viewed in the eyepiece of the instrument.

3.3 Mass

The method of measuring mass is with a *balance*. In fact, balances compare the *weight* of the unknown mass with the weight of a standard mass. But because weight is proportional to mass, equality between the unknown weight and the weight of the standard mass means that the unknown mass is equal to the standard mass.

In your laboratory, you may have access to a number of different types of balance, including the top-pan balance, the lever balance and the spring balance. It is important that you should familiarise yourself with the use of all types that are available to you, so that you do not restrict your choice to one particular type. Note also that some types of spring balance may be calibrated in force units (that is, in newtons) rather than in mass units (kilograms).

The top-pan balance

The top-pan balance is a direct-reading instrument, based on a pressure sensor. The unknown mass is placed on the pan, and its weight applies a force to the sensor. The mass corresponding to this force is displayed on a digital read-out.

When using the balance, ensure that the initial (unloaded) reading is zero. There is a control for adjusting the zero reading. The balance may have a tare facility, allowing you to set the scale to zero while an empty container is on the balance. This means that the mass of material added to the container is obtained directly. This works in the same way as adjusting the balance for zero error.

Top-pan balance

The uncertainty in the reading of a particular top-pan balance will be quoted in the manufacturer's manual. As with other digital instruments, it is likely to be expressed as a percentage uncertainty of the reading shown on the scale, together with the uncertainty in the final figure of the display (see page 7).

The spring balance and the lever balance

Spring balances are based on Hooke's law: the extension of a loaded spring is proportional to the load. The extension is measured directly, by a marker moving along a straight scale, or by a pointer moving over a circular scale. As with any instrument using a scale and pointer, you should take care not to introduce a parallax error when you take readings. Position yourself so that your line of sight is perpendicular to the scale. Before placing the object of unknown mass on the pan, check for zero error. There is likely to be a zero-error adjustment screw on the balance.

Spring balance

Lever balances are based on the principle of moments. In one common type, the unknown mass is placed on a pan, and balance is achieved by sliding a mass along a bar, calibrated in mass units, until the bar is horizontal. This represents the condition that the moment of the load is equal and opposite to the moment of the sliding mass on the bar. A reading is taken from the edge of the sliding mass of the divisions marked on the bar. In this case, parallax error is less likely to be serious. Again, check for zero error before taking a reading.

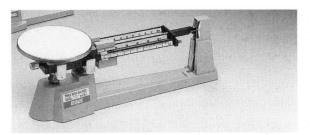

Lever balance (horizontal)

Lever balance (circular)

Another type of lever balance has a pointer moving along a circular scale. A weight on the pointer arm is placed in one of two positions in order to change the range (for example, from 0 – 100 g to 0 – 1 kg).

Both of these types of balance are used more for the convenience of obtaining a rapid, approximate reading, rather than for an accurate determination. (They can be regarded as being the kitchen scales of the laboratory.) An indication of the uncertainty involved in readings with a particular balance can be obtained from the smallest division on the scale.

Example

The mass of a quantity of chemical is determined using a lever balance. Over the range of masses involved, the separation between mass graduations on the bar is 2 g. The mass of the empty container is 56 g, and the mass of the container plus the chemical is 104 g. Find the mass of the chemical, and the uncertainty in this value.

By subtraction, the mass of the chemical is 104 – 56 = **48 g**. The uncertainty in each reading is likely to be the difference between mass graduations on the beam, that is ± 2 g. Each of the two readings has an uncertainty of ± 2 g: the uncertainty in the mass of the chemical is thus **± 4 g**.

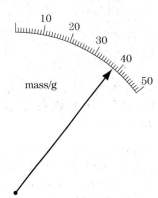

Figure 3.14a

Now it's your turn

The mass of a chemical used to make up a solution is determined as follows. A dish containing the chemical is placed on the pan of a spring balance. The pointer reading on the scale is shown in Figure 3.14a. The chemical is then tipped into a known volume of water, and the empty dish replaced on the pan, giving the pointer reading shown in Figure 3.14b. What is the mass of chemical, and what is the uncertainty in this value?

Ans: 15 g; ± 2 g

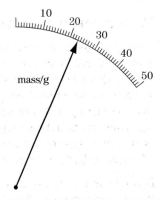

Figure 3.14b

Choice of method

As stated above, the top-pan and spring balance are direct-reading instruments. This means that readings can be obtained quickly and conveniently. The lever balance requires adjustment of the sliding mass, but this takes only a very short time. There may be some house rules in your laboratory about which type of balance should be used for which task. In general, the pan of the balance should be kept clean and dry. Do not weigh out loose chemicals on the pan; always use a container, the mass of which you have determined beforehand, or for which you have made allowance using the tare control.

In general, choose a balance of sensitivity appropriate to the experiment you are carrying out.

Application: current balance

A U-shaped magnet is placed on a top-pan balance (Figure 3.15). A wire is clamped so that it runs along the channel of the magnet. The wire is connected in a circuit with a d.c. supply, a rheostat, an ammeter and a switch. When the supply is switched on, the balance reading is seen to change, because a force is exerted on the wire in the magnetic field. By Newton's third law, a force is also exerted on the magnet. This is detected by the change Δm in the mass reading. This change must be converted into force F by multiplying Δm by g. The variation with current I of the magnetic force F may be determined, and the form of the equation:

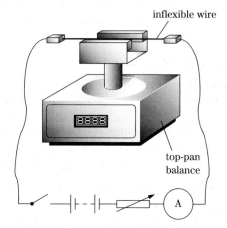

Figure 3.15 Current balance experiment

$$F = BIl$$

(where l is the length of the wire in the magnetic field) may be verified. The direction of the force, as predicted theoretically by the Left-Hand Rule, may also be verified by checking whether the mass reading increases or decreases for a given current direction relative to the magnetic field.

3.4 Time

The experiments you will meet in your practical Physics course deal with the measurement of time intervals, rather than with absolute time. The basic method of measuring a time interval is with a stopclock or stopwatch. In each case, the instrument is started and stopped by pressing a lever or a button, and reset by pressing another control. You should familiarise yourself with the method of operating the instrument before you start a timing experiment in earnest. Bear in mind that the reaction time of the experimenter (a few tenths of a second) is likely to be much greater than the uncertainty of the instrument itself. If steps are not taken to reduce the effects of reaction time, an unacceptable systematic error may be built in to the experiment. As has been mentioned on page 12, one way of reducing the effect of reaction time is to time enough events (for example, the swings of a pendulum) to make the interval being measured very much longer than the

Analogue stopclock

experimenter's reaction time. A good technique is to count the events (the swings), commencing by counting down to zero, and starting the timer at the zero count. Wherever possible, work with at least twenty seconds' worth of events (oscillations), and repeat each set of timings three times. (Sometimes, when carrying out experiments on damped oscillations, you will have to be satisfied with fewer swings, but try not to go below intervals of ten seconds.)

Digital stopclock

The stopclock

A mechanical, spring-powered stopclock will have an analogue display, that is, a hand (or hands) which move round a dial. Such an instrument is likely to read to the nearest one-fifth of a second.

The stopwatch

This instrument is likely to have a digital display. It is based on the oscillations of a quartz crystal. The read-out will probably be to the nearest one-hundredth of a second. In addition to the start, stop and reset controls, digital stopwatches often have a 'lap' facility, which allows one reading to be held in the display while the watch is still running. Because of this complexity, it is vital that you know the functions of all the controls.

Your own wristwatch may well have a built-in stopwatch, which will be no less precise than the watches available in the laboratory. However, the start and stop controls on wrist stopwatches are sometimes rather small, and it is important that you should not fumble a start or stop signal.

Choice of method

Often, students are attracted to a digital stopwatch because it reads to one-hundredth of a second. However, in all experiments in which the start and stop signals are applied manually, such accuracy is overkill. The reaction time of the experimenter, which is likely to be a few tenths of a second, will greatly outweigh the accuracy of the watch. It would be misleading, and bad practice, to enter times such as '21.112 s' in a table of results if the systematic error due to reaction time had not been fully accounted for. Thus, if you are doing an experiment on the timing of oscillations and your classmates have acquired all the digital timers which are available, you will be at no disadvantage if you have to use an analogue stopclock.

Application: determination of the acceleration of free fall

A steel sphere is released from an electromagnet and falls under gravity. As it falls, it passes through light gates which switch an electronic timer on and off (Figure 3.16). The acceleration of free fall can be determined from the values of the time intervals and distances.

This is an experiment in which electronic switching is essential in order to reduce the potentially very large error caused by the reaction time of the experimenter. Here timing to one-hundredth of a second is essential.

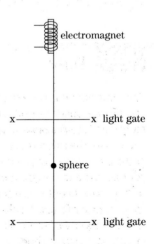

Figure 3.16 *Acceleration of free-fall*

Application: measurement of frequency using a cathode-ray oscilloscope

A cathode-ray oscilloscope (c.r.o.) has a calibrated timebase, so that measurements from the screen of the c.r.o. can be used to give values of time intervals. One application is to measure the frequency of a periodic signal, for example the sine-wave output of a signal generator. The signal is connected to the Y-input of the c.r.o., and the Y-amplifier and timebase controls are adjusted until a trace of at least one, but less than about five, complete cycles of the signal is obtained on the screen. The distance L on the graticule (the scale on the screen) corresponding to one complete cycle is measured (Figure 3.17). It is good practice to measure the length of, say, four cycles, and then divide by four so as to obtain an average value of L.

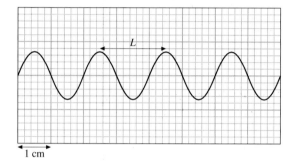

Figure 3.17 *Measurement of frequency*

The graticule will probably be divided into centimetre and perhaps millimetre or two-millimetre divisions. If the timebase setting is x (which will be in units of seconds, milliseconds or microseconds per centimetre) the time T for one cycle is given by $T = Lx$. The frequency f of the signal is then obtained from $f = 1/T$. The uncertainty of the determination will depend on how well you can estimate the measurement of the length of the cycle from the graticule. Bearing in mind that the trace has a finite width, you can probably measure this length to an uncertainty of about ± 2 mm.

As with most instruments you will use, your laboratory time will be so limited that you will probably have to take the timebase settings on trust. However, it is worth thinking about possible methods of checking the calibration. You could try checking against a calibrated signal generator: but who is to say which of the signal generator or c.r.o. has the correct calibration? Another method would be to connect a microphone to the Y-input, and sound a tuning fork of known frequency near the microphone.

Example

The output of a signal generator is connected to the Y-input of a c.r.o. When the timebase control is set at 0.50 milliseconds per centimetre, the trace shown in Figure 3.18 is obtained. What is the frequency of the signal?

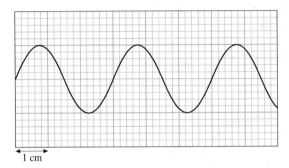

1 cm

Figure 3.18

Two complete cycles of the trace occupy 6.0 cm on the graticule. The length of one cycle is therefore 3.0 cm. The timebase setting is 0.50 ms cm^{-1}, so 3.0 cm is equivalent to $3.0 \times 0.50 = 1.5$ ms. The frequency is thus $1/1.5 \times 10^{-3} =$ **670 Hz**.

Now it's your turn

The same signal is applied to the Y-input of the c.r.o. as in the example above, but the timebase control is changed to 2.0 milliseconds per centimetre. How many complete cycles of the trace will appear on the screen (which is 8.0 cm wide)?

Ans: 10

3.5 Temperature

The SI unit of temperature, the kelvin, is based on the ideal gas (or thermodynamic) scale of temperature. The scale may be arrived at using an instrument called a constant-volume gas thermometer. The equation relating the Celsius temperature scale to the thermodynamic scale is

$$t = T - 273.15$$

where t is in degrees Celsius and T is in kelvin.

Fortunately, in your practical course you will come across nothing more complicated than a liquid-in-glass (probably a mercury-in-glass) thermometer. You may also do experiments on thermocouple thermometers and resistance thermometers, and assess whether a thermistor would be suitable for use as a thermometer.

The mercury-in-glass thermometer

Liquid-in-glass thermometers are based on the thermal expansion of a liquid. A quantity of liquid is contained in a bulb at the end of a thin capillary tube. The space above the liquid contains an inert gas at low pressure or a vacuum. If the bulb is placed in a beaker of water which is gradually heated, the liquid expands and the thread of liquid occupies more and more of the capillary. The capillary is graduated: the position of the end of the thread gives the temperature (Figure 3.19).

Most thermal physics experiments which you will carry out will involve the measurement of temperatures between 0 °C (the temperature of melting ice) and 100 °C (the temperature of steam above boiling water at a pressure of 1 atmosphere). The most useful thermometer covering this range is a mercury-in-glass thermometer with graduations from –10 °C to 110 °C, in 1 °C intervals. You will find it easy to take readings to the nearest half degree, and perhaps to 0.2 °C. There are a number of precautions you should take when using the thermometer. If you are measuring the temperature of a beaker of liquid which is being heated, the liquid must be thoroughly stirred before taking the reading. (Because of convection currents, there is a temperature difference of several degrees between the top and the bottom of the liquid.) The thermometer is calibrated for use at a standard depth of immersion; this may be stated on the stem. If it is not, try to ensure that the thermometer is always immersed in the liquid to the same depth. The length of the bulb plus about 20 mm is a good rule of thumb.

There are some points to be made about safety. Thermometers are relatively fragile instruments. Because of their shape, they have a tendency to roll along the bench-top. Make sure that your thermometer does not roll off and fall to the ground. If a thermometer does break and the mercury in it comes out, do not be tempted to play with the mercury. Mercury is a poison. To reduce the risk of breakage, do not use the thermometer as a stirrer, unless it is of a robust type designated as a 'stirring thermometer'. If you have to fit a rubber bung to a thermometer, make sure that the hole in the bung is large enough, and lubricate the rubber thoroughly with soap. Wear gloves and grip the thermometer so that, if it breaks, your wrist will not be cut.

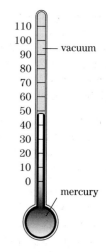

Figure 3.19 *Mercury-in-glass thermometer*

Example

The temperature of a mixture of ice, salt and water is measured using a mercury-in-glass thermometer. When thermal equilibrium has been reached, the mercury thread in the thermometer is as shown in Figure 3.20. What is the temperature of the mixture? What is the uncertainty in this value?

By interpolation between the scale divisions, the temperature reading is **–2.5** °C. The uncertainty is probably about ± **0.2** °C.

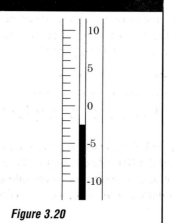

Figure 3.20

Now it's your turn

The temperature of a solidifying liquid is measured using a liquid-in-glass thermometer. When thermal equilibrium has been reached, the liquid thread in the thermometer is as shown in Figure 3.21. What is the solification temperature? What is the uncertainty in this value?

Ans: 37.8 °C; about ± 0.2 °C

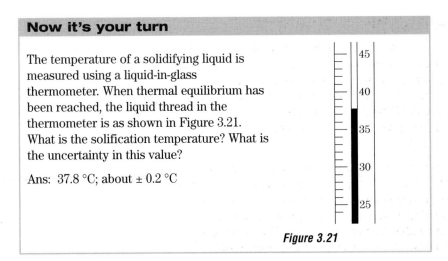

Figure 3.21

The thermocouple thermometer

A thermocouple thermometer consists of two wires made of different metals or alloys, joined at one end. The other ends of the wires are connected to the terminals of a millivoltmeter. This may be a digital instrument, which is calibrated in °C. The thermocouple may also be connected to a data-logger. The junction is placed in thermal contact with the object, the temperature of which is required.

The thermocouple thermometer actually measures the difference in temperature between the junction of the two metals (the hot junction) and a cold junction. In some applications, the cold junction is placed in an ice-water mixture, so as to achieve a known reference.

Choice of method

The heat capacity of the bulb of a liquid-in-glass thermometer is much greater than that of the hot junction of a thermocouple. For this reason, the thermocouple is particularly useful if a rapidly-varying temperature is to be measured, or if the object, the temperature of which is required, has a small thermal capacity. Mercury-in-glass thermometers are available to cover the temperature range from about –40 °C to 350 °C. Thermocouples using different pairs of metal or alloy wires can cover a much larger range.

The choice of a particular thermometer in a given application will depend on the range of temperatures to be covered, the thermal capacity of the object, and whether the temperature is varying rapidly.

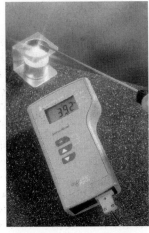

Thermocouple thermometer

3.6 Current and potential difference

Your Physics laboratory will probably have a selection of instruments for measuring current and potential difference (voltage). The two main types are *analogue* meters, in which a pointer moves over a scale, and *digital,* in which the value is displayed on a read-out consisting of a series of integers.

CHAPTER 3

Analogue meters

The normal analogue meter is restricted to the measurement of the relevant quantity over a single range. For example, a 0 – 1 A d.c. ammeter will measure direct currents in the range from zero to 1 A. A 0 – 30 V d.c. voltmeter will measure steady potential differences in the range from zero to 30 V. Some analogue meters have a dual-range facility, with a common negative terminal and two positive terminals, each of which is associated with a separate scale on the instrument. Thus, one scale might be 0 – 3 A, and the other 0 – 10 A. Each of the positive terminals is marked with the scale to which it refers. Be careful to take the reading on the scale corresponding to the pair of terminals you have selected.

Analogue ammeter

Analogue meters are subject to zero error. Before switching on the circuit, check whether the needle is exactly at the zero graduation. If it is not, return the needle to zero by adjusting the screw at the needle pivot. There is also the possibility of parallax error. The needle should be read from a position directly above it and the scale, and not from one side. Sometimes a strip of mirror is provided close to the scale so that the experimenter can align the needle with its image in the mirror, ensuring that viewing is vertical. The uncertainty associated with a current or voltage reading from an analogue meter is usually taken to be ± half the smallest scale reading.

A *galvanometer* is a sensitive current-measuring analogue meter. A galvanometer may be converted into an ammeter by the connection of a suitable resistor in parallel with the meter (Figure 3.22). Such a resistor is called a *shunt*. The meter may be converted into a voltmeter by the connection of a suitable resistor in series with the meter (Figure 3.23). Such a resistor is called a *multiplier*. In some theory examinations, the calculation of the values of shunts and multipliers in order to give desired current or voltage ranges is a popular type of question. Fortunately, in the laboratory, you do not need to do these sums. The manufacturers provide shunts and multipliers which are clearly labelled with the conversion function and full-scale deflection, for attachment to the basic galvanometer. All you need do is to select the shunt or multiplier required for your experiment and make sure that you apply the correct factor when reading the scale.

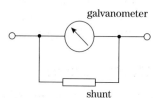

Figure 3.22 *Galvanometer with shunt, for current measurements*

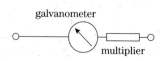

Figure 3.23 *Galvanometer with multiplier for voltage measurements*

Shunts and multipliers for use with a galvanometer

Digital meters

The use of a digital meter may save you the trouble of selecting an instrument with the right range for your application. Most have an auto-ranging function, that is, the instrument selects the most sensitive range for the particular value of current or voltage being measured. All the experimenter has to do is to check whether there is a zero error and adjust if necessary, to note whether the display indicates 'A' or 'mA', and to observe the position of the decimal point.

As mentioned in Section 1.2, the uncertainty in the reading of a digital meter is expressed in terms of the overall uncertainty and the uncertainty in the last digit. When in use, you will note that the last digit of the display fluctuates from one figure to another. You can try to estimate the mean of the fluctuations, but if this fluctuation occurs, there is clearly uncertainty in the last digit of the value.

Digital ammeter

Multimeters

Multimeters, or multifunction instruments, are available in both analogue and digital forms. Such meters may include switched options for the measurement of direct and alternating currents and voltages, and of resistance, with several ranges for each quantity being measured. If you use a multimeter, make sure that you are familiar with the controls, so that you can set the instrument to measure the quantity you require.

Choice of method

Much will depend on the selection of meters available in your laboratory. Before you set up your circuit, make a rough calculation to determine the ranges of currents and voltages that you will have to measure. This is a vital part of the planning process, and will help to make sure that you select the appropriate instrument from those that are available.

Digital multimeter

In some laboratories, multimeters are provided for use primarily as test instruments, which are meant to be available to anyone who wishes to make a rapid check on currents, potential differences or resistances in a circuit. If this is the rule in your laboratory, it is bad practice to tie up a multimeter in a long experiment, when a single-function and single-range instrument would do the job equally successfully. However, in other laboratories, sufficient multimeters may be available for students not to feel guilty about using them.

Remember that, to measure a current in a component in a circuit, an ammeter should be connected in series with the component. To measure the potential difference across the component, a voltmeter should be connected in parallel with the component. The arrangement is shown in Figure 3.24.

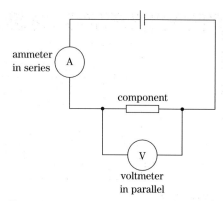

Figure 3.24 An ammeter is connected in series with the component, a voltmeter in parallel

Application: measurement of voltage using a cathode-ray oscilloscope

The cathode-ray oscilloscope, with its calibrated Y-amplifier, may be used to measure the amplitude of an alternating voltage signal. (We have already seen, in Section 3.4, how the timebase of the c.r.o. may be used to measure time.) The signal is connected to the Y-input, and the Y-amplifier and timebase settings are adjusted until a suitable trace is obtained (Figure 3.25). The amplitude A of the trace is measured. If the Y-amplifier setting is Q (in units of volts per centimetre), the peak value V_0 of the signal is given by $V_0 = AQ$. The peak-to-peak value is $2V_0$, and the r.m.s. voltage is $V_0/\sqrt{2}$. Remember that the reading obtained on an analogue or digital voltmeter is the r.m.s. value.

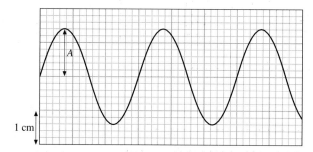

Figure 3.25 Measurement of alternating current

Example

The output from a signal generator is connected to the Y-input of a c.r.o. When the Y-amplifier control is set to 5.0 millivolts per centimetre, the trace shown in Figure 3.26 is obtained. Find (a) the peak voltage of the signal, (b) the r.m.s. voltage.

Example *continued*

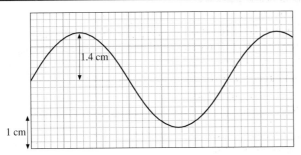

Figure 3.26

Measure the amplitude of the trace on the graticule: this is 1.4 cm. The Y-amplifier setting is 5.0 mV cm⁻¹. 1.4 cm is thus equivalent to $1.4 \times 5.0 = 7.0$ mV. The peak voltage of the signal is **7.0 mV**. The r.m.s. voltage is given by $7.0/\sqrt{2} =$ **4.9 mV r.m.s.**

Now it's your turn

The output from a signal generator is connected to the Y-input of a c.r.o. When the Y-amplifier control is set to 20 millivolts per centimetre, the trace shown in Figure 3.27 is obtained. Find (a) the peak-to-peak voltage of the signal, (b) the r.m.s. voltage.

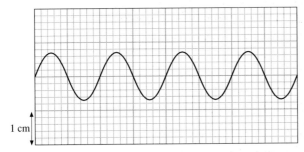

Figure 3.27

Ans: **28 mV; 9.9 mV r.m.s.**

3.7 Planning an experiment

In Sections 3.2 to 3.6, we have looked at methods of measuring the base quantities of length, mass, time, temperature and current. In your practical course, you will be assessed on your planning ability. This starts with testing whether you know which instrument is the most suitable for the required measurement. Should you use a metre rule, vernier calipers, micrometer screw gauge or travelling microscope to measure the diameter of a steel sphere? What type of thermometer should you use in an experiment involving the electrical heating of liquid in a container? What voltage and current ranges should you select for the meters in an experiment to measure resistance by the voltmeter-ammeter method? This chapter should have provided the answers to this sort of problem of choice.

You may also be required to carry out a full planning and design exercise. Very often this will not involve the use of any apparatus, nor will you have to take any readings. However, you may be expected to call on your knowledge of theory topics which are within your syllabus.

In such an exercise, you should identify all the possible variables in the investigation. Normally, the instructions given to you will make it clear what these variables are likely to be. You should then consider all the variables to decide which quantities to vary, and which to keep constant. Remember to work with only one pair of variables at a time.

You should then think about the experimental methods and techniques which are likely to be available to you. In some types of planning questions, a list of apparatus (which will certainly include several items which are quite unsuitable, and which are there only as red herrings) will be provided. In others, you will be invited to use 'any facilities of your school or college laboratory'. Always consider more than one experimental technique, and come to a conclusion about the method likely to be most suitable. In your answer, explain the factors you have considered in making your choice.

Depending on the structure of the question, you may then be expected to write an account of the procedure for the experiment. This may possibly be in the form of a set of instructions for a fellow-student, or an account of what you intend to do yourself. You should illustrate this section with diagrams, and make clear how the measurements are to be made.

You may also be asked how the readings would be processed to obtain the required result. For example, you should explain what graph you would plot, and how the result could be obtained from the graph.

Finally, you should remember that the quality of your written communication may be assessed on this type of question. For this reason, it is important that you should write clearly and concisely, using correct grammar, spelling and punctuation.

Chapter 3 Summary

★ Methods available for the measurement of length include:

- metre rule (range 1 m, reading uncertainty 0.5 mm)
- micrometer screw gauge (range 50 mm, reading uncertainty 0.01 mm)
- vernier caliper (range 100 mm, reading uncertainty 0.1 mm)
- travelling microscope (range 250 mm, reading uncertainty 0.1 mm)

★ Methods available for the measurement of mass include:

- top-pan balance
- spring balance
- lever balance

★ Methods available for the measurement of time include:

- stopclock (reading uncertainty 0.2 s)
- stopwatch (reading uncertainty 0.01 s)
- cathode-ray oscilloscope

Note that the reaction time of the experimenter is likely to be a few tenths of a second.

★ Methods available for the measurement of temperature include

- liquid-in-glass thermometers
- thermocouple thermometers

★ Methods available for the measurement of current and potential difference include:

- analogue meters
- digital meters
- multimeters
- cathode-ray oscilloscope

CHAPTER 3

Chapter 3 Questions

1   You are asked to measure the internal diameter of a glass capillary tube (diameter about 2 mm). You are also to investigate the uniformity of the tube along its length. Suggest suitable methods.

2   When using a travelling microscope, students are warned to beware of the possibility of the screw mechanism introducing an error due to backlash. Explain how this type of error may arise, and what precautions should be taken against it. A micrometer screw gauge also employs a screw mechanism. Discuss whether or not it is necessary to take precautions against backlash in this case.

3   (a) The value of the acceleration of free fall varies slightly at different places on the Earth's surface. Discuss whether the following should be calibrated when moved to different locations:
    (i)   a top-pan balance
    (ii) a spring balance
    (iii) a lever balance
    (b) How would you calibrate a balance?

4   The shutter on a particular camera has settings which allow it to be open for (nominally) 1 s, 0.5 s, 0.25 s, 0.125 s, 0.067 s, 0.033 s, 0.017 s, 0.008 s, 0.004 s, 0.002 s and 0.001 s. Suggest a method (or methods) of calibrating the exposure times over this range.

5   Explain the factors you would consider when deciding whether to use a liquid-in-glass or a thermoelectric thermometer in particular experimental situations.

6   Summarise the advantages and disadvantages of analogue and digital ammeters.

7   Explain how to use a cathode-ray oscilloscope to measure the characteristics of the sinusoidal output from a signal generator.

CHAPTER FOUR

Evaluation and communication

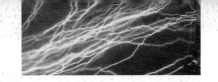

The aim of this chapter is to develop the skill of evaluation. When we evaluate an experimental result, we make an assessment of its reliability. We come to a decision about whether the experiment has proved or disproved a hypothesis, or we make a critical comparison of a numerical result which has just been obtained with a value in a reference book.

Throughout Chapter 3, the uncertainties likely to be encountered if you measure length, mass, time, temperature or current using a particular technique or instrument have been emphasised. Now we shall see how these uncertainties are combined to give an indication of the overall uncertainty in the final result.

In many experiments, the results are presented in the form of a straight-line graph. We shall also see how to evaluate results obtained from the gradient or intercept of such a graph.

Another aspect of evaluation is the critical examination of the design of the experiment. Has the experiment achieved its aim? If not, are there any procedures which could be improved? Which is the vital measurement in the experiment? Have you made the best use of laboratory resources and of your own time?

In many cases, you will find it easier to evaluate your experiment through discussion with fellow-students and teachers. To make the most of such sessions, you will need to develop your communication skills in order to present your case clearly. Hand-waving arguments and gut feelings will not be sufficient. You will need to back up your arguments with numbers and graphs.

Communication skills are vital in this aspect of your assessment. Whether you are taking a written practical examination or following teacher-assessed coursework, you will need to show that you can plan, design and evaluate experiments. A substantial proportion of the marks for Experiment and Investigation is reserved for this aspect. A student who cannot communicate ideas will be at an enormous disadvantage in this part of the test.

4.1 Combining uncertainties

In this section, two simple rules for obtaining an estimate of the overall uncertainty in a final result will be stated.

The rules are:

1 For quantities which are added or subtracted to give a final result, add the actual uncertainties.

2 For quantities which are multiplied together or divided to give a final result, add the fractional uncertainties.

Suppose that we wish to obtain the value of a physical quantity x by measuring two other quantities y and z. The relation between x, y and z is known, and is

$$x = y + z$$

If the uncertainties in y and z are Δy and Δz respectively, the uncertainty Δx in x is given by

$$\Delta x = \Delta y + \Delta z$$

If the quantity x is given by

$$x = y - z$$

the uncertainty in x is again given by

$$\Delta x = \Delta y + \Delta z$$

Example

1 I_1 and I_2 are two currents coming into a junction in a circuit. The current I going out of the junction is given by

$$I = I_1 + I_2$$

In an experiment, the values of I_1 and I_2 are determined as 2.0 ± 0.1 A and 1.5 ± 0.2 A respectively. What is the value of I? What is the uncertainty in this value?

Using the given equation, the value of I is given by $I = 2.0 + 1.5$ = 3.5 A. The rule for combining the uncertainties gives $\Delta I = 0.1 + 0.2$ = 0.3 A. The result for I is thus 3.5 ± 0.3 A.

2 In an experiment, a liquid is heated electrically, causing the temperature to change from 20.0 ± 0.2 °C to 21.5 ± 0.5 °C. Find the change of temperature, with its associated uncertainty.

The change of temperature is $21.5 - 20.0 = 1.5$ °C.
The rule for combining the uncertainties gives the uncertainty in the temperature change as $0.2 + 0.5 = 0.7$ °C.
The result for the temperature change is thus **1.5 ± 0.7 °C**.

The second example shows that a small difference between two quantities may have a large uncertainty, even if the uncertainty in measuring each of the quantities is small. This is an important factor in considering the design of experiments, where the difference between two quantities may introduce an unacceptably large error.

Now it's your turn

A travelling microscope is used to measure the diameter of a capillary tube. Readings on the vernier scale of the microscope at opposite ends of a diameter are 24.15 mm and 22.92 mm. Each reading has an uncertainty of ± 0.01 mm. What is the diameter of the tube? What is the uncertainty in the diameter?

Ans: 1.23 ± 0.02 mm

There is a different rule for combining the uncertainties when quantities are multiplied together, raised to a power, or divided to give the final result. Suppose a quantity x is given by

$$x = A \, y^a \, z^b$$

where A is a constant.

The uncertainty in the measurement of y is $\pm \Delta y$, and that in z is $\pm \Delta z$. Expressed as a fraction of y, the fractional uncertainty in y is $\pm \Delta y/y$. Similarly, the fractional uncertainty in z is $\pm \Delta z/z$. The fractional uncertainty in x is given by the following rule:

$$\Delta x/x = a(\Delta y/y) + b(\Delta z/z)$$

Example

A value of the acceleration of free fall g was determined by measuring the period of oscillation T of a simple pendulum of length l. The relation between g, T and l is

$$g = 4\pi^2(l/T^2)$$

In the experiment, l was measured as 0.55 ± 0.02 m, and T was measured as 1.50 ± 0.02 s. Find the value of g, and the uncertainty in this value.

Substituting in the equation, $g = 4\pi^2(0.55/1.50^2) = $ **9.7 m s^{-2}**. The fractional uncertainties are $\Delta l/l = 0.02/0.55 = 0.036$ and $\Delta T/T = 0.02/1.50 = 0.013$. Applying the rule to find the fractional uncertainty in g, $\Delta g/g = \Delta l/l + 2\Delta T/T = 0.036 + 2 \times 0.013 = 0.062$. The actual uncertainty in g is given by (value of g) × (fractional uncertainty in g) = $9.7 \times 0.062 = 0.60$ m s^{-2}. The experimental value of g, with its uncertainty, is thus **(9.7 ± 0.6) m s^{-2}**. (Incidentally, it is not good practice to determine g from the measurement of the period of a pendulum of fixed length. It would be much better to take values of T for a number of different lengths l, and to draw a graph of T^2 against l. The gradient of this graph is $4\pi^2/g$.)

Now it's your turn

A value of the volume V of a cylinder is determined by measuring the radius r and the length L. The relation between V, r and L is

$$V = \pi r^2 L$$

In an experiment, r was measured as 3.30 ± 0.05 cm, and l was measured as 25.4 ± 0.4 cm. Find the value of V, and the uncertainty in this value.

Ans: (870 ± 40) cm^3

If you find it difficult to deal with the fractional uncertainty rule, you can easily estimate the uncertainty by substituting extreme values into the equation. Taking account of the uncertainties in y and z, the lowest value of x is given by

$$x_{low} = A(y - \Delta y)^a(z - \Delta z)^b$$

and the highest by

$$x_{high} = A(y + \Delta y)^a(z + \Delta z)^b$$

If x_{low} and x_{high} are worked out, the uncertainty in the value of x is given by $(x_{high} - x_{low})/2$.

Example

Apply the extreme value method to the data for the simple pendulum experiment.

Because of the form of the equation for g, the lowest value for g will be obtained if the lowest value of l and the highest value for T are substituted. This gives

$$g_{low} = 4\pi^2(0.53/1.52^2) = 9.1 \text{ m s}^{-2}$$

The highest value for g is obtained by substituting the highest value for l and the lowest value for T. This gives

$$g_{high} = 4\pi^2(0.57/1.48^2) = 10.3 \text{ m s}^{-2}$$

The uncertainty in the value of g is thus $(g_{high} - g_{low})/2 = (10.3 - 9.1)/2 =$ **0.6 m s^{-2}**, as before.

Now it's your turn

Apply the extreme value method to the data for the volume of the cylinder.

If the expression for the quantity under consideration involves both products (or quotients) and sums (or differences), then the best method of attack is the extreme value method.

4.2 The evaluation of graphs

Many experiments are designed to provide data which can be plotted so as to yield a straight line graph. The way in which information can be obtained from the gradient and intercept has been discussed in Section 2.2.

Suppose that a simple theory suggests that two quantities x and y are connected by the relation $y = mx + c$. Theoretically, if pairs of values of x and y are obtained and plotted on a graph of y (vertical axis) against x (horizontal axis), a straight line of gradient m and intercept c on the vertical axis should be obtained. Experimentally, the result of plotting the points may not give a perfect straight line. Figure 4.1 shows a graph in which the points are so scattered that there is no indication of the expected straight-line relationship. In this case, we would say that x and y are *uncorrelated* (that is, there is no relationship between x and y), and the conclusion is that the simple theory is wrong. This is an extreme example, and it is unlikely that you will come across such a negative result. You will usually be steered to graphs which show a correlation between the two quantities. In such cases, there may be a degree of scatter, but there is clearly an indication of a straight-line graph. Figure 4.2 illustrates a graph in which an experimenter might feel that the relation between x and y is indeed linear. However, the points do not lie exactly on a straight line, and we need to know how to draw what is described as the *best fit line* through them. Details of how to choose the scales for a graph, plot the points, draw the best fit line, and determine the gradient and intercept have already been given in Section 2.3. We will return to the procedure at the stage of finding the best fit line and evaluating the quantities (the gradient and intercept) which can be obtained from the graph.

The first step is to locate what may be called the centre of mass (or *centroid*) of the plotted points. If the points are spread evenly along the ranges of measurement of the two axes, then the centroid will be about half-way along each of these ranges. In Figure 4.3, the centroid of the points plotted in Figure 4.2 has been marked C. The best straight line will pass through C.

Now take a ruler (a transparent one is best, so that you can see the plotted points through it) and try to locate a line which, when it passes through C, will have an equal number of points above and below it. Furthermore, the sum of the distances from the line of the points on one side of the line should be approximately equal to the sum of the distances from the line of the points on the other side of the line. This line gives an estimate of the best fit line. The procedure has been carried out in Figure 4.4.

We can now determine the gradient and intercept of the best fit line, as explained on pages 27 and 28.

The next stage is to try to estimate the position of two limiting lines of good fit (as distinct from best fit). This procedure has been carried out in Figure 4.5. Note that the two limiting lines are not lines of the lowest and highest possible gradients for the set of points (this would give lines which could not possibly be regarded as lines of good fit). You may feel dissatisfied

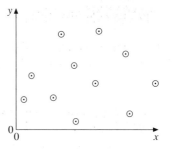

Figure 4.1

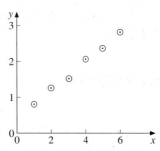

Figure 4.2

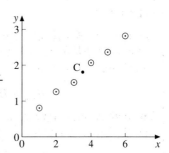

Figure 4.3

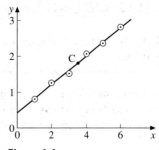

Figure 4.4

because subjective judgment is becoming important here, but remember that all uncertainties are based on estimates. Now that we have the two limiting lines, we can find the range of values in which the gradient and intercept can lie, and can quote uncertainties for each.

If you wish to shorten this procedure, you can estimate the position of only one of the limiting lines of good fit, say the steeper one. Values of the gradient and the intercept for this line can be determined. The uncertainty in the gradient is then the difference between the gradient of the best fit line and the gradient of the good fit line. The uncertainty in the intercept is the difference between the intercept of the best fit line and the intercept of the good fit line.

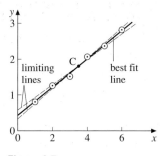

Figure 4.5

As stated above, this procedure leads to rather rough estimates of uncertainty. However, there is a quantitative procedure for calculating the uncertainties in gradient and intercept. What is called a *least-squares line* is the best straight line through the set of plotted points, based on the condition that the sum of the squares of the deviations of the points from the line should be a minimum. You may have an electronic calculator which offers a program for calculating the least- squares line. However, if this refinement is not available, the computation involved in calculating a least-squares line is quite tedious, and to obtain the uncertainties in the least-squares gradient and intercept is even more time-consuming. This quantitative treatment of the best fit line is not required in your AS/A2 course. Indeed, you will not gain credit for merely plugging numbers into the calculator and using the least-squares program. It is much more important to show that you understand what is involved in making an estimate of the uncertainty in the gradient. The best way of demonstrating your understanding of the procedure is to sketch an extreme line, and find by how much its gradient and intercept differ from that of the best fit line.

Example

In an experiment to determine the acceleration of free fall g, the period T of a simple pendulum was determined for various values of the length l. The following results were obtained:

l/m	0.25	0.35	0.45	0.55	0.65	0.75
T/s	0.98	1.21	1.34	1.50	1.60	1.75

(Note that this is only part of the complete table of results, which would show that for each length, the time for a certain number of swings – perhaps 20 – was determined. Repeat values of these time readings would also be recorded.)

The relation between T, l and g is

$$T^2 = 4\pi^2(l/g)$$

Plot a graph of T^2 against l, and use it to obtain a value for g, with the associated uncertainty.

Example *continued*

The values of T^2 corresponding to the various values of l are first tabulated.

l/m	0.25	0.35	0.45	0.55	0.65	0.75
T^2/s^2	0.96	1.46	1.80	2.25	2.56	3.06

Figure 4.6 shows the graph of T^2 against l. Now locate the centroid C of the plotted points. Because the values of l and T^2 are evenly-spaced, C will be close to the middle of the range of values of l and T^2. C is marked on Figure 4.6: by calculation (although this is not essential), you can show that it is at l/m = 0.50 and T^2/s^2 = 2.02. Now try to draw the best fit line. This, too, is indicated on Figure 4.6. The gradient of this line is (3.22 − 0.02)/0.80 = 4.00 s^2 m^{-1}. There is an intercept of 0.02 on the T^2/s^{2}axis.

To obtain the value of g, remember that the equation of the line is

$$T^2 = 4\pi^2(l/g).$$

Thus, the gradient of the graph of T^2 against l is $4\pi^2/g$. Substituting tthe value for the gradient, we have $g = 4\pi^2/4.00 = 9.9$ m s^{-2}.

Now estimate the position of one of the two limiting lines of good fit. The steeper limiting line is shown on Figure 4.6. Because the scatter of the points is small, the limiting line is very close to the best fit line; it is easy to confuse the two lines. It is a good idea to distinguish the limiting line in some way. The gradient of the limiting line in Figure 4.6 is 4.13 s^2 m^{-1} . The corresponding value of g is 9.6 m s^{-2}. The value of g, with the associated uncertainty, is thus **(9.9 ± 0.2) m s^{-2}**. Note that the uncertainty is quoted to one significant figure. The answer is then given to one decimal place to correspond with the uncertainty.

Should we be worried about the fact that the best fit line has a small positive intercept, when the theoretical equation says that the line should go through the origin? No: if two limiting lines, the steeper and the shallower, are drawn, they clearly include the origin as a possible point on the graph.

Now it's your turn

In an experiment to determine the resistance R of a resistor, values of the potential difference V across the resistor were determined for various values of the current I through it. The following results were obtained:

I/mA	10	20	30	40	50	60
V/V	0.20	0.46	0.62	0.90	1.08	1.36

(Note that this is only part of the complete table of results.)

The relation between I, V and R is

$$V = IR$$

Use the data to draw a suitable graph, and find the value of R, with the associated uncertainty.

Ans: 22 ± 1 Ω

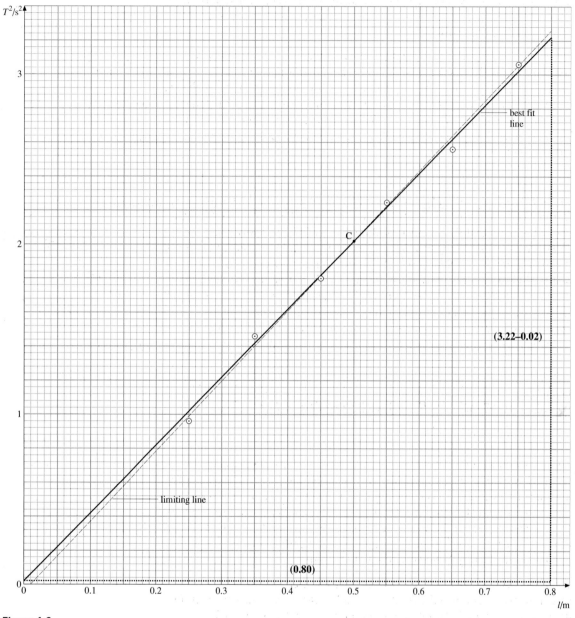

Figure 4.6

In Section 2.2 we have learnt how to convert mathematical functions, such as relations involving powers or the exponential function, into the straight-line form. This is achieved by taking logarithms of both sides of the equation. This means that the procedure we have just gone through (and which has been illustrated with reference to very familiar experiments) can be applied to many other physical situations. We can use the idea of drawing limiting lines on a graph to put limits on the values of exponents in unknown relationships, or on decay constants in radioactive decay, or on time constants in experiments involving the charge and discharge of capacitors through resistors.

A word of warning. Because you have been trained to deal primarily with straight-line graphs, it is easy to assume that all variations you are investigating experimentally obey a linear relationship. This may not always be true. Figure 4.7a shows a graph of the rate of flow Q of a liquid along a tube as a function of the pressure difference P between the ends of the tube. The theory of steady (streamline) flow suggests that the relation between Q and P should be of the form

$$Q = KP$$

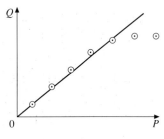

Figure 4.7a

where K is a constant. That is, a graph of Q against P should be a straight line through the origin. The experimenter may feel that the points are close enough to a straight line to claim that the theory holds. However, at high flow-rates the flow of liquid is not steady (it is said to be turbulent), and the graph of Q against P will flatten off. To determine whether the last few points of Figure 4.7a were obtained under steady or non-steady conditions, it is worthwhile to include error bars indicating the uncertainty in individual readings. This procedure has been indicated in Figure 4.7b. If the uncertainty in individual readings is as low as shown there, then it seems likely that the transition to non-steady flow has indeed taken place.

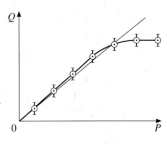

Figure 4.7b

4.3 The evaluation of experimental procedures

Besides evaluating your graphs and experimental results, you will be required to assess whether your experimental procedures are adequate for the task. To do this, you will need to have an appreciation of possible sources of systematic uncertainty for the instruments you have used. Pages 10–13 give a general introduction to systematic uncertainty, and throughout Chapter 3, details were given of the sources of uncertainty likely to be encountered in using different types of instrument.

You also need to assess the importance of random error. If you have displayed your results in the form of a graph, a large random error will show up as an unacceptably large scatter of points. You will need to make suggestions about how the situation can be improved. Remember, the effects of random error can be reduced by taking a larger number of results.

The evaluation of experimental procedures will certainly be expected of you in relation to experiments and investigations you have actually carried out. The evaluation may also occur in the context of planning and design.

Although, as a starting point, you should list the types of measurement taken, together with the uncertainty in each, this is not the whole task. You need to look at the way the measurements are put together in obtaining the final result. Very often this will lead you to realise that one of the quantities to be measured is the critical quantity. You may then have to make special efforts to measure that quantity as accurately as possible, whereas you can afford to be less precise in measuring others. This can be illustrated with reference to an example taken from a planning/design situation.

Example

Under conditions of steady flow, the rate of flow Q of liquid along a horizontal capillary tube of radius r and length L depends on the pressure difference Δp between the ends of the tube. The relation between these quantities is

$$Q = A \, \Delta p \, r^4/L$$

where A is a constant which depends on the liquid involved. Evaluate the experimental procedures which might be used in an experiment to determine the value of A.

Without going into full details about the design of this experiment, it is likely that you would suggest that Q could be measured by timing the delivery of a known volume of liquid, using a measuring cylinder and stopclock. Δp could be measured using a manometer with a metre rule (or a travelling microscope if the pressure difference was very small). L could be measured with a metre rule. What about r? Because r appears in the equation raised to the fourth power, the final uncertainty in Q is likely to be dominated by the uncertainty in r. Remember that for expressions of this form the contribution of uncertainties in individual quantities depends on the power to which they are raised. Thus, the fractional uncertainty in A is given by the fractional uncertainty in Q, plus the fractional uncertainty in Δp, plus the fractional uncertainty in L, plus *four times* the fractional uncertainty in r! You should therefore go to some trouble to measure r as accurately as you can. You would probably use a travelling microscope to find the diameter of the tube, and if the tube could be cut up after the experiment, take readings at a number of sections along the tube so that you could check for uniformity of the bore and obtain a reliable average value of r.

Now it's your turn

In an experiment to investigate the motion of a sphere falling through a liquid, a metal sphere of radius r is released from rest at the surface of oil in a tall, vertical container. The sphere accelerates initially, but eventually moves with a constant velocity v_t (the terminal velocity). The relation between v_t and r is

$$v_t = B \, r^2 \, g \, (\sigma - \rho)$$

where B is a constant which depends on the liquid involved, and σ and ρ are the densities of the metal and the oil respectively. Evaluate the experimental procedures which might be used in an experiment to determine the value of B.

In your evaluation of experimental procedures, you should try to suggest possible improvements within the general framework of the existing experiment. At this level, the suggestion that the whole experiment should be scrapped, and a completely new method tried, would be regarded as too radical a solution.

4.4 Communicating your work

How much communication you have to do in the course of your assessment under the Experiment and Investigation aspect of your AS/A2 Specification will depend on whether you are working towards a formal practical examination, or whether your coursework is being assessed by your teacher. In the case of the practical examination, you will communicate only through a written examination paper. In coursework, both written accounts and presentation/interview sessions will be involved. In either case, you will need to be able to describe your work, state your results and present your conclusions in an appropriate way.

If your experiment is intended to demonstrate a physical phenomenon, describe the observations clearly. Make use of sketches and diagrams if this helps to make the description more clear.

If your experiment is intended to prove or disprove a hypothesis, state the result clearly and unambiguously. Statements like 'this was a good experiment' are too vague. It is not clear if the experiment was considered to be good because the student enjoyed doing it, or because new techniques and skills were learnt, or because the result proved the hypothesis.

If your experiment is intended to provide a numerical value for a physical quantity, then that value, together with its associated uncertainty and unit, must be stated clearly. Numerical values and uncertainties must be quoted to an appropriate number of significant figures. If the quantity is one which can be compared with a value from a reference book, then quote the reference value also. If there is a discrepancy between your experimental result and the reference value, your evaluation discussion should point to possible explanations. Finally, avoid statements like 'correct to within experimental error', which are almost meaningless.

Chapter 4 Summary

★ Combining uncertainties:

- in expressions of the form $x = y + x$, $\Delta x = \Delta y + \Delta z$

- in expressions of the form $x = Ay^a x^b$,
  $\Delta x/x = a(\Delta y/y) + b(\Delta z/z)$

★ To evaluate a linear graph,

- locate the centroid of the plotted points

- draw a best fit line through the centroid and the points

- determine the gradient and intercept of the best fit line

- estimate a limiting line of good fit

- determine the uncertainty in the gradient and intercept from the limiting line and the best fit line.

★ When evaluating experimental procedures,

- list the types of measurement and the uncertainty in each

- look at the scatter of points on a graph

- identify the critical quantity in the determination

- consider alternative, more appropriate approaches within the general framework of the experiment

★ When communicating experimental work,

- describe observations clearly, using sketches and diagrams

- state whether a hypothesis has been proved or disproved

- state your numerical value for the physical quantity determined, together with its uncertainty and unit; quote also the value from a reference book.

CHAPTER FIVE

Assessment and tasks

This Chapter is concerned with the assessment of the experimental skills which you have been practising during your physics course. We shall be considering various experimental tasks which may be set to test these skills.

A word of warning – although you may feel good if you complete the assessment of some of the skills early in your course, this could be a mistake. You need to practise the skills. Early assessment may mean that you have not achieved your full potential before assessment.

Some sections of this Chapter may appear to have been written with the teacher in mind. Indeed, this is the case. Unless a task has been set correctly, then it is not possible for the student to achieve full credit. However, it is useful for students to know why and how tasks are being set in order that they may reach their full potential.

5.1 Assessment of Experimental Skills

In Chapters 1 to 4, we looked at the skills of practical work that should be developed during an Advanced level physics course. All of these skills will be assessed at both AS and at Advanced levels. Broadly speaking, the assessment is made under four headings.

Skill 1: Planning
For this skill you should be able to identify and define a problem using information given to you on a task sheet and from your knowledge of physics. You should be able to choose effective and safe procedures, and describe suitable apparatus and methods to solve the given problem.

Skill 2: Implementing
For this skill a particular experiment or task will be outlined for you. You should be able to set up apparatus efficiently and use it safely and effectively. You should be able to make and record observations and measurements bearing in mind the precision of the apparatus provided and, if appropriate, use IT. Also, you should be able to identify sources of systematic and random error and to modify procedures so that the results you obtain are as accurate and reliable as possible.

Skill 3: Analysing evidence and drawing conclusions
For this skill you should be able to analyse observations and present processed data in appropriate ways. This may be, for example, as part of a table, or in graphical form. You should be aware of the limitations of the measurements you have taken when commenting on trends or patterns in the data. You are also expected to draw valid conclusions using your knowledge and understanding of physics.

CHAPTER 5

Skill 4: Evaluating evidence and procedures

This skill is concerned with the assessment of the reliability of data and the conclusions drawn from them. You are expected to show an awareness of the limitations imposed by the apparatus on any result obtained.

You should appreciate that Skill 3 is concerned with the data and observations, and the processing of the data. Skill 4 deals with the conclusions based on the processed data.

Throughout all experimental work, safety considerations are of paramount importance. Aspects of safety are assessed in Skills 1 and 2 and it is here that students are expected to make statements regarding safety and to carry out procedures with due regard for safety. It must be remembered that, although students must be aware of their own safety and should carry out procedures safely, the ultimate responsibility for safety does rest, in the first instance, with the teacher in the laboratory.

Some publications commonly used as a basis for safety in the school/college laboratory are given below. Although these publications are aimed primarily at teachers, they may be of interest to students, particularly for Skills 1 and 2 since these Skills contain the assessment of 'safe practice'.

Safety in Science Education, DfEE, 1996, HMSO, ISBN 0 11270915 X

Safeguards in the School Laboratory, 10th Edition, 1996, ASE, ISBN 0 86357 250 2

Hazcards, 1995, CLEAPSS School Science Service

Laboratory Handbook, 1998-97, CLEAPSS School Science Service

Topics in Safety, 2nd Edition, 1988, ASE ISBN 0 86357 104 2

Safety Reprints, 1996 Edition, ASE ISBN 0 86357 246 4

Hazardous Chemicals, A Manual for Science Education, SSERC Limited 1997, ISBN 0 95317760 2

Experimental skills may be assessed either by means of an examination or through coursework. Regardless of the means of assessment, all four skills will be used in the assessment. It is important that you should practise each skill using a series of tasks. Any individual task may concentrate on any one skill or on more than one skill.

In this chapter, some tasks will be outlined which may be used to test individual skills or more than one skill. A list of suitable experiments for the testing of the skills will also be given.

Task 1

Assessment of Skill 1: Planning

One task suitable for the assessment of Skill 1 is an investigation of the bounce of a table-tennis ball. A task sheet for this investigation is given opposite.

The bounce of a table-tennis ball

You are to plan an experiment to determine how a certain factor which you will choose affects the bounce of a table tennis ball.

In your plan you should:

1 consider what is meant by the bounce of a ball and how it may be measured scientifically,

2 discuss factors which affect this bounce,

3 choose one factor to be investigated,

4. outline different procedures for the chosen investigation and give reasons for the procedure you decide to use,

5 describe the apparatus to be used and how to make the necessary measurements

6. show clearly how you will process the measurements in order to determine the relation you are investigating.

Notes for Teachers

When setting a planning task, the investigation should be such that it does not make unnecessary demands on apparatus in the laboratory but, at the same time, provides an opportunity for candidates to demonstrate their skills at Advanced level. The task should have a number of variables and students must be aware that variables, other than the dependent and independent variables, must be kept constant.

Ideally, after students have written their plans, they should be given the opportunity to try out their experiments so that they can make appropriate modifications to their plans before they are submitted for assessment.

The task sheet for the student gives a statement of the task to be conducted together with a brief outline of the contents. It is important to note that this outline gives no detail as to what is required in the task. For example, candidates are not told that they should include in the outline of the experiment means by which unwanted variables are kept constant. If details such as these are given, then the candidate cannot be assessed on that particular aspect and, consequently, a low mark for the candidate would result.

Task 2

Assessment of Skill 2: Implementing

An example of a task suitable for the assessment of Skill 2 is given below.

The magnification of the image produced by a convex lens

The image of an object produced by a convex lens may not be the same size as the object.

The magnification m of the image of the object is given by the expression

$$m = \text{length of image} / \text{length of object}$$

In this experiment you are to investigate how the magnification m depends on the distance d between the centre of the lens and the image itself.

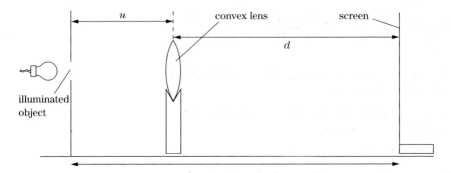

Figure 5.1

You are provided with an object consisting of a metal sheet in the centre of which has been punched a hole. A wire is stretched across the hole. A lamp may be used to illuminate the object. The image is observed on a graph-paper screen.

1. Measure and record the diameter of the hole which is being used as the object.

2. Set up the apparatus as shown in Figure 5.1, with the centre of the lens on the same horizontal level as the centre of the object. Adjust the position of the screen and lens so that the distance d between the centre of the lens and the screen is about 30 cm.

3. Change the position of the object so that a clear image of the object is formed by the lens on the screen.

4. Measure and record the values of d and the diameter of the image.

5. Repeat the procedure for values of d in the range $d = 30$ cm to $d = 80$ cm.

6. Determine the value of the magnification m for each value of d.

7. Plot a graph of m (y-axis) against d (x-axis).

8. Hence determine the relation between m and d.

Notes for Teachers

This task is to assess Skill 2, that is, the setting up of apparatus, the making and recording of observations and the identification of errors. Candidates should adapt procedures to reduce such errors. Instructions 1 and 5 are sufficient to assess this Skill. The remaining instructions have been included for completeness of the task involved and would not form part of the assessment.

Task 3

Assessment of Skill 3: Analysing evidence and drawing conclusions

Absorption of β-radiation in aluminium

β-radiation from a radioactive source is known to be absorbed in aluminium. In this experiment you are to investigate the variation with thickness of aluminium of the transmitted radiation.

WARNING: This experiment involves the use of a radioactive source. The following safety precautions must be adhered to at all times.

- The radioactive source is to be moved only by the teacher. You must ask the teacher to make any adjustments of the source.

- Do not place any part of your body in front of the source.

- When placing or removing sheets of aluminium from above the source, use the long tweezers provided, keeping your hands well away from the source.

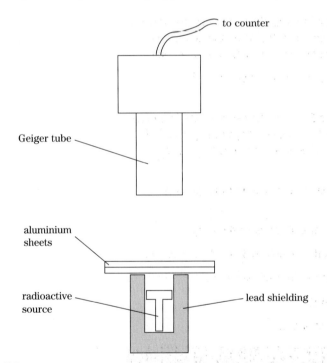

Figure 5.2

1. Set up the Geiger tube and counter as shown in Figure 5.2, without the radioactive source in place.

   The source in its shielding must be several metres away from the Geiger tube.

2. Switch on the counter. The counter should be counting randomly at about 20 counts per minute. If this is not the case, consult your teacher. Record the number of counts in 5 minutes (300 s).

3. Switch off the counter.

4. Ask your teacher to place the radioactive source, in its lead shielding but with the top removed, below the geiger tube.

5. Measure and record the thickness of one of the sheets of aluminium and place this on top of the shielding, as shown in Figure 5.2. Do not move the source or the Geiger tube during the remainder of the experiment.

6. Switch on the counter and record the number of counts in 300 s. Switch off the counter.

7. Measure and record the thickness of another sheet of aluminium. Place between this and the first sheet and the Geiger tube so that the thickness of the absorber is now increased.

8. Repeat instructions 6 and 7 until the total thickness of absorber is about 1.5 cm.

9. The variation of the count rate C from a β-emitting radioactive source is thought to vary with the thickness of absorber x according to the relation

$$C = ae^{-bx}$$

Use your experimental data to determine whether this relation is followed.

10. Draw conclusions as to the relation involved, indicating how limitations of the measurements taken affect the conclusions.

Notes for Teachers

Skill 3 is not concerned with data collection and recording. The recording and use made of processed data is of importance here.

One problem which some candidates will experience is that the quality of their raw data is such that they are handicapped when attempting to process this data. For example, if the raw data is not reliable, then any graphical work is likely to result in a wide scatter of points with consequent low reliability of any conclusions. Since the data collection is not part of the assessment, the experiment could be conducted as a group demonstration. It is important that students do experience the data collection, otherwise they are not able to comment from first-hand knowledge of limitations in measurement.

Task 4

Assessment of Skill 4: Evaluating evidence and procedures
The oscillation of a suspended magnet

1. Set up the apparatus as shown in Figure 5.3.

    The rule is to be horizontal and the supporting threads for the suspended Magnadur magnet are to be of equal length.

2. Stick the second Magnadur magnet to the bench with plasticene so that the two magnets attract one another and the suspended magnet is vertically above the fixed magnet.

3. Adjust the height of the suspended magnet so that the distance x between its lower face and the upper face of the fixed magnet is 5.0 cm. Measure and record x.

4. Displace the suspended magnet so that it makes small oscillations in a direction perpendicular to the plane containing the supporting threads and the rule.

5. Make and record measurements to determine the period T of these oscillations.

6. Repeat instructions 4 and 5 for values of x in the range 3.5 cm to 6.0 cm, until you have six sets of readings of x and the corresponding values of T.

7. Determine lg T and lg x and plot a graph of lg T (y-axis) against lg x (x-axis).

8. The quantities T and x are related by the expression

$$T = ax^n$$

Use your graph to determine the values of a and n.

9. Determine a value for T for a separation x of 2.5 cm.

10. Using your knowledge of physics and your experience of carrying out this investigation, assess the reliability of the data and of the conclusions drawn from them. Also discuss the limitations imposed by the apparatus on the processed data.

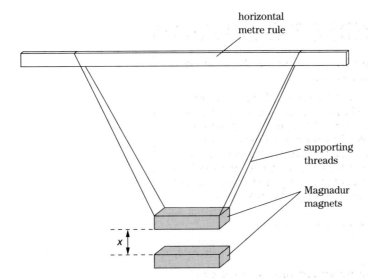

horizontal
metre rule

supporting
threads

Magnadur
magnets

x

Figure 5.3

Notes for Teachers

It should be noticed that the assessment of Skill 4 is almost entirely encompassed within instruction 10 of the task. The remainder of the experiment is concerned with giving the student the necessary experience so that instruction 10 may be fully answered.

The assessment of more than one Skill in a given task

It can be seen that in order to assess Skill 3, the student must have experience of the experimental procedures and data collection. These experiences are included within Skill 2. Similarly, for the assessment of Skill 4, students must have first-hand experience of aspects of Skills 2 and 3 in that particular task. This does beg the question as to whether more than one Skill can be assessed within one task. The answer must be 'yes'.

As an example, consider the task set for the assessment of Skill 4 (see page 70). It was intended to give sufficient detailed instruction so that Skill 4 could be assessed effectively. To this end, the data must be of sufficient quality and quantity and the processing must be adequately accomplished. Consequently, too much detail in the instructions has been given for the assessment of Skills 2 and 3.

For the assessment of Skill 2, no reference should be made to the number of sets of data. Furthermore, for Skill 3, instruction 7 and any reference to drawing a graph in instruction 8 would have to be deleted. It is possible, therefore, to assess Skills 2, 3 and 4 using the task on page 70 and 71, but this may not be advisable.

It can be seen that the assessment of the Skills is in some ways sequential. Poor data collection would adversely affect any assessment of Skills 3 and 4. Similarly, poor processing would adversely affect performance in Skill 4. The decision as to whether to use a task for multiple assessment does, therefore, lie with the teacher. The more highly skilled the student as regards experimental techniques, then the more likely it is that multiple assessment will be effective in providing the student with an optimum mark in each Skill.

In Section 5.1 it was stated that some of the better planning activities are those in which students try out their ideas in the laboratory. Obviously, such a practice allows the student to improve on the planning. It does mean that the student can use the plan to carry out a follow-on experiment for the assessment of Skills 2, 3 and 4. Again, the advisability of such a practice has to be considered carefully. It is unlikely that all criteria for all Skills can be met effectively on one task.

An example of a task designed to test more than one Skill is outlined below.

Temperature characteristic of a thermistor

The resistance R of a thermistor is thought to vary with thermodynamic temperature T according to the expression

$$R = R_0 \, e^{b/T},$$

where R_0 and b are constants.

You are to determine corresponding values of R/Ω and T/K and hence verify the relation, determining the values of R_0 and b.

You are provided with the following equipment.

> thermistor with long insulated leads
> thermometer, -10 °C to 110 °C
> ammeter
> voltmeter
> battery or d.c. power supply
> switch
> variable resistor
> connecting wires
> several beakers
> ice water and ice
> hot water
> stirrer

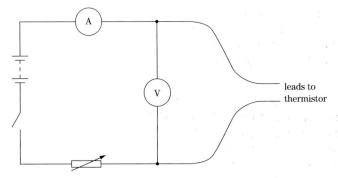

Figure 5.4

You may ask for any other items which you think are necessary.

1. Set up the circuit as shown in Figure 5.4.

2. Arrange the thermistor and the thermometer in a water bath so that the thermometer may be used to measure the temperature of the thermistor.

3. Fill the water bath with ice and water.

4. Measure and record the temperature T of the thermistor, the voltmeter reading V and the ammeter reading I.

5. Determine R, the resistance of the thermistor, where $R = V/I$.

6. Repeat the procedure to determine corresponding values of R and T for values of T in the range $T = 0$ °C to $T = 40$ °C.

7. Show that the relation

$$R = R_0 \, e^{b/T}$$

is valid, and determine the values of b and R_0.

8. Comment fully on the limitations of your measurements and the validity of your results. Also consider the limitations of the apparatus used.

Notes for Teachers

In order to assess Skill 1, instructions 1 to 7 should be omitted. These instructions do in fact give an outline of the plan for the experiment.

The task is suitable for the assessment of Skill 2, which includes safety considerations and the efficient handling of the equipment. These aspects can best be recorded by the supervisor during the course of the experiment by means of a tick list. The list provides documentary evidence for later moderation of the work. For this particular task, the tick list could include:

Efficient handling of equipment:

- circuit set up correctly without assistance
- thermistor placed close to thermometer
- thermistor and thermometer held appropriately in clamp on boss and stand
- appropriate action (e.g. avoiding parallax) when reading meters and thermometer
- time allowed for thermistor, thermometer and water to come into equilibrium
- water transfers made efficiently.

Safety:

- apparatus set up in appropriate position to avoid accidents
- circuit switched off while making connections
- circuit switched off between readings
- initially, variable resistor set at maximum
- use of a stirrer, not a thermometer, to stir water
- no water spilled or, if spillage occurs, water is mopped up immediately.

The level of detail given to students as regards the measurements to be taken allows the following to be assessed in Skill 2:

- appropriate number of readings
- appropriate range/spacing of readings
- repetition of readings
- appropriate tabulation of readings
- readings given to an appropriate number of significant figures.

Skill 3 may also be assessed. The task allows candidates to:

- process data giving appropriate numbers of significant figures
- develop the theory so that an appropriate graph may be plotted
- process data by plotting a suitable graph
- draw appropriate conclusions from graphical analysis.

continued

Notes for Teachers *continued*

Skill 4 may be assessed by consideration of the appropriateness of methods adopted and the validity of the results. Candidates should:

- comment on procedures adopted and how they attempted to improve the precision/accuracy of their measurements

- comment on the reliability of their results and, where appropriate, estimate quantitatively the reliability of the findings

- comment on shortcomings in apparatus and techniques and how improvements may be made.

5.2 Tasks for the assessment of experimental skills

It can be seen on page 74 that any particular task may be tailored to the assessment of one or more Skills. Consequently, it is not intended that this Section should give specific details of assessment of Skills but rather, a number of suitable tasks will be outlined. Task sheets can then be written, tailored to suit the Skill to be tested and to meet the needs of students.

Acceleration of a ball down a slope

For a ball moving down a ramp of length L and height h (see Figure 5.5) the linear acceleration a of the ball down the slope is given by

$$a = 2s/t^2$$

where s is the distance moved from rest down the ramp in time t.

Students may

1 investigate the dependence of s on t,

2 plot a graph of s against t^2 to determine a.

Note: At first sight, it may be thought that a is related to the acceleration of free fall g by the expression

$$a = g \sin \theta$$

where
$$\sin \theta = h/L$$

Of course, the relation does not take into account rotation of the ball. It is a useful exercise for students to determine $a/\sin \theta$, and then to argue as to why this value is so far from the accepted value for g.

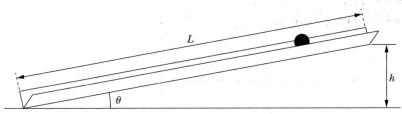

Figure 5.5

CHAPTER 5

Trajectory motion

When an object is projected horizontally near the Earth's surface, the horizontal distance x through which the ball travels is dependent on the height h through which it falls (see Figure 5.6).

The distances y and x are related by the expression

$$y = gx^2/2v^2$$

where g is the acceleration of free fall and v is the horizontal component of velocity.

The variation with x of y may be investigated with the arrangement illustrated in Figure 5.7.

The ball rolls down the ramp and, when it hits the board, a mark is left on the paper, enabling the distance x to be measured.

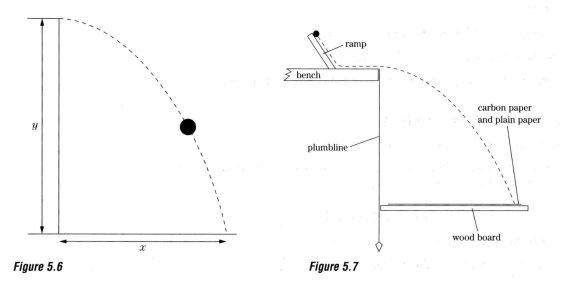

Figure 5.6

Figure 5.7

The cantilever

A cantilever, loaded with a mass M, is illustrated in Figure 5.8.

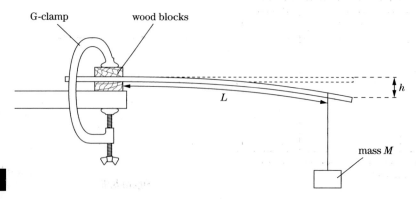

Figure 5.8

The depression h of the free end of the cantilever depends on the weight of the mass. The degree of bending will also depend on the width b and the thickness d of the cantilever, together with the Young modulus E of the material from which it is made. The depression h is given by the expression

$$h = (4MgL^3)/Ebd^3$$

where g is the acceleration of free fall and L is the effective length of the cantilever.

Students may

1   investigate the variation with L of the depression h,

2   determine the Young modulus E.

The period T of small vertical oscillations of the loaded cantilever is given by the expression

$$T = 2\pi \{(4ML^3)/(bd^3E)\}^{\frac{1}{2}}$$

Students may

1   investigate the variation with L of the period T,

2   investigate the variation with M of the period T,

3   determine the Young modulus E of the material of the cantilever.

The bifilar suspension

A bifilar suspension is illustrated in Figure 5.9.

The period T of small oscillations about a vertical axis through the centre of the pendulum is given by the expression

$$T = 2\pi \sqrt{(kL/d^2)}$$

where L is the length of the supporting threads,

         d is the separation of the parallel threads

and k is a constant for the pendulum.

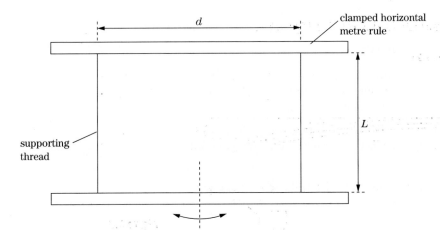

Figure 5.9

For a pendulum where the supporting threads are not parallel, the period T of small oscillations becomes

$$T = 2\pi \sqrt{(kL / d_1 d_2)}$$

where d_1 and d_2 are the separation of the supporting threads at the upper and lower points of suspension respectively.

Students may investigate how the period T of oscillation depends on

1 the length L, keeping d constant,

2 the separation d of parallel threads, keeping L constant,

3 the separation d_1, keeping L and d_2 constant,

4 the separation d_2, keeping L and d_1 constant.

The helical spring

The time period T of vertical oscillations of a helical spring loaded with a mass m is given by the expression

$$T = 2\pi \sqrt{(k/m)}$$

where k is the elastic constant of the spring.

The weight mg of the mass causes an extension e of the spring and, assuming the elastic limit of the spring has not been exceeded,

$$k = mg/e$$

Substituting, $T = 2\pi \sqrt{(g/e)}$

Students may

1 investigate the dependence of T on the suspended mass m,

2 investigate the dependence of T on the extension,

3 determine a value for the acceleration of free fall g.

If two or more similar springs are available, then the relation of the elastic constant for springs connected in series and in parallel (see Figure 5.10) to the elastic constant of a single spring may be investigated.

The compound pendulum

A uniform rod (e.g. a metre rule) is drilled at regular intervals along its length. The rod is pivoted freely at one hole and the rod is set into small amplitude oscillations in a vertical plane, as illustrated in Figure 5.11.

The period T of oscillation is found for different distances d of the pivot from one end of the rod. Plotting T against d, a graph similar to Figure 5.12 will be obtained.

A horizontal line drawn at any particular value of T will cut the graph lines in four places. The lengths l are determined for various values of T.

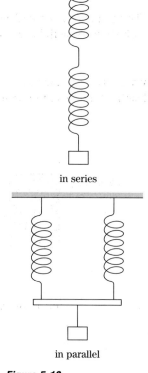

in series

in parallel

Figure 5.10

T and l are related by the expression

$$g = 4\pi^2 \, l/T^2$$

where g is the acceleration of free fall.

Students may

1   investigate the relation between T and l,

2   determine the acceleration of free fall.

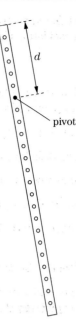

Figure 5.11

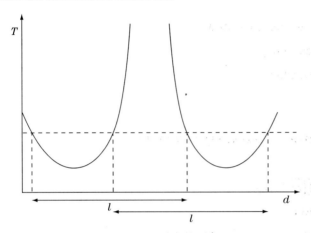

Figure 5.12

Liquid flow

The flow of water through a horizontal capillary tube may be investigated using the apparatus of Figure 5.13.

The volume Q of water flowing per unit time through a horizontal tube of radius r and length l is given by the expression

$$Q = p\pi r^4/8l\eta$$

where η is the viscosity of water and p is the pressure difference across the length of the tube. For water of density ρ,

$$p = h\rho g$$

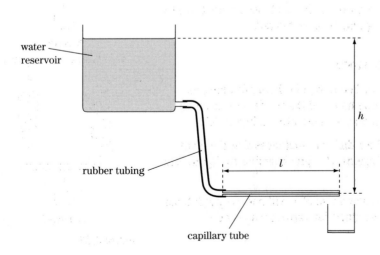

capillary tube

Figure 5.13

where g is the acceleration of free fall and h is the pressure head. Combining,

$$Q = h\rho g\pi r^4/8l\eta$$

If the volume of water in the reservoir is much greater than the volume which flows through the tube during the experiment, the pressure head may be assumed to be constant. Note that the viscosity of water is temperature dependent.

Students may investigate the dependence of flow-rate Q on

1 h, keeping r and l constant,

2 r, keeping h and l constant,

3 l, keeping r and h constant,

4 water temperature.

The flow of water from a vertical burette may be used to investigate the variation with time of the length l of the water column in the burette. The variation will be of the form

$$l = l_0 e^{-bt}$$

where l_0 and b are constants.

Students may verify the form of the relation and determine the constants.

Newton's law of cooling

The law states that, under conditions of forced convection, the rate of loss of heat from a body is proportional to the excess temperature of the body above its surroundings.

Apparatus is set up as shown in Figure 5.14.

A cooling curve for the hot container is obtained, as illustrated in Figure 5.15.

The gradient of this curve (the rate of cooling H) is found at various excess temperatures t and a graph of H (y-axis) is plotted against t (x-axis).

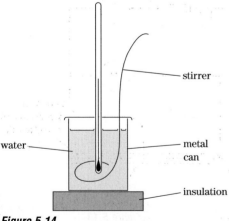

stirrer

water

metal can

insulation

Figure 5.14

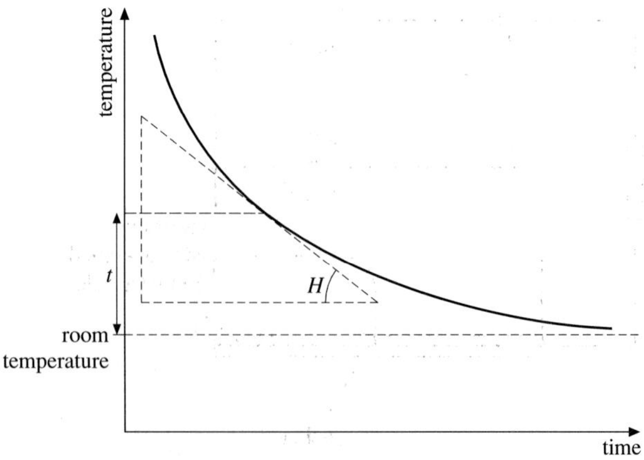

Figure 5.15

Students may then make deductions from this graph.

The experiment may be repeated under conditions of natural convection.

Power transfer of a battery

A bare resistance wire of resistance per unit length about $10 \ \Omega \ m^{-1}$ is stretched along a metre rule. The wire is connected into the circuit of Figure 5.16.

For various lengths l of wire, the readings of the ammeter I and the voltmeter V are taken. Plotting a graph of resistance R ($= V/I$) against length l enables the wire to be calibrated.

A battery of e.m.f. about 3 V is connected in series with a resistor of resistance about 3 Ω. The purpose of the resistor is to provide the battery with appreciable 'internal' resistance. The battery is connected into the circuit of Figure 5.17.

For various lengths l of wire, the readings of the ammeter I and the voltmeter V are taken. The power output P of the battery can be determined ($= VI$). A graph is then plotted of P (y-axis) against resistance R (x-axis) of the wire. The value of R for maximum power transfer can be obtained.

If resistance wires of different diameters are available, a calibration graph for each wire enables an investigation of the effect of diameter of the wire on resistance to be carried out.

If students are informed that maximum power is transferred when the external resistance is equal to the internal resistance, then the internal resistance of the battery can be determined. What may be more profitable is to change the 'internal' resistance of the battery and then to show the relation between external resistance and 'internal' resistance.

The experiment can be used to investigate change of internal resistance as a battery runs down. Determination of the charge delivered by the battery serves as a measure of the life of the battery.

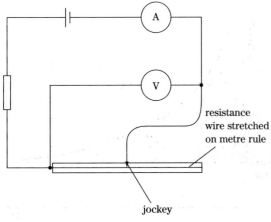

resistance
wire stretched
on metre rule

jockey

resistance
wire stretched
on metre rule

jockey

Figure 5.16

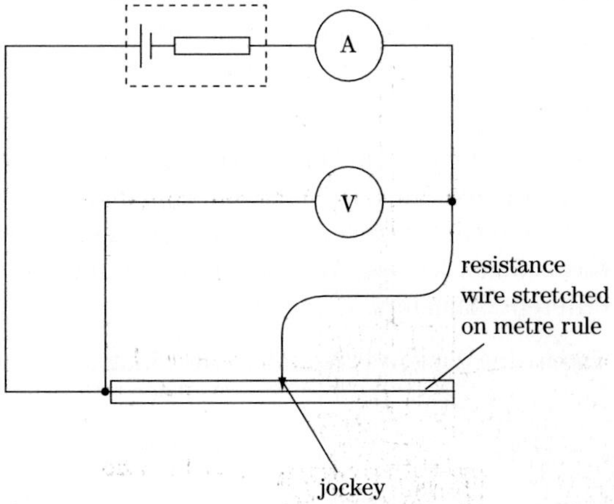

Figure 5.17

CHAPTER 5

Discharge of a capacitor

A capacitor of capacitance C and a resistor of resistance R are connected into the circuit of Figure 5.18.

The time constant CR of the circuit should be about 200 s and the voltmeter must have a resistance of the order of 10^{10} Ω (a digital voltmeter is suitable).

The potential difference V across the resistor is found at different times t.

It is known that the variation with time t of the potential difference V is given by

$$V = V_0 \, e^{-t/CR}$$

The student may

1 investigate the discharge of a capacitor by plotting a graph of ln V against t.

2 determine resistance if the value of capacitance is known,

3 determine capacitance if the value of resistance is known,

4 determine the resistance of a moving-coil voltmeter.

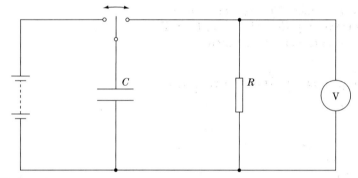

Figure 5.18

Fusing current

The variation with diameter of a wire of the current required to fuse (melt) the wire can be investigated using the apparatus of Figure 5.19.

The diameter d of one of the wires is measured and it is then placed in the circuit. The current is gradually increased until the wire fuses and the fusing current I is noted – take suitable safety precautions as this is done. The experiment is repeated for a number of wires of different diameters.

The fusing current is related to the diameter by an expression of the form

$$I = kd^n$$

Plotting a graph of lg I against lg d allows this expression to be investigated.

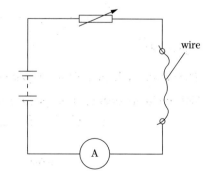

Figure 5.19

Note that the length of wire is not a factor in the expression. Nevertheless, fusing current will be affected if the wire is very short . Students may wish to investigate whether length is a factor.

Focal length of a convex lens

This experiment can provide useful practice of graphical analysis.

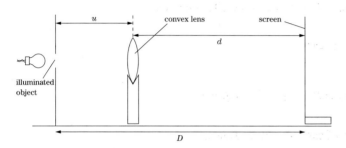

Figure 5.20

The apparatus is illustrated in Figure 5.20.

A clear image of the object is focussed on the screen. The distance u from the centre of the lens to the object is measured, together with the distance D between the object and the screen. The measurements are repeated for different values of u.

A graph of D (y-axis) against u (x-axis) is plotted. This graph is shown in Figure 5.21.

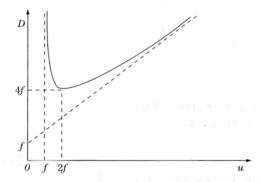

Figure 5.21

The curve is a hyperbola having asymptotes $u = f$ and $D = u + f$.

The minimum occurs at the point $u = 2f$, $D = 4f$.

CHAPTER 5

Group demonstration experiments

As discussed on page 70, some group experiments may well be useful for the assessment of Skills 3 and 4. These experiments include:

- determination of e/m_e for electrons using a fine-beam tube
- determination of e/m_e for electrons using a Teltron tube
- determination of the Planck constant using the photoelectric effect
- absorption of γ-radiation in aluminium
- absorption of β- and γ-radiation from a mixed source in aluminium
- determination of the half-life of radon.

INDEX

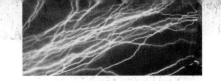